INTRODUCTION TO NONLINEAR
AND GLOBAL OPTIMIZATION

Springer Optimization and Its Applications

VOLUME 37

Managing Editor
Panos M. Pardalos (University of Florida)

Editor — Combinatorial Optimization
Ding-Zhu Du (University of Texas at Dallas)

Advisory Board
J. Birge (University of Chicago)
C.A. Floudas (Princeton University)
F. Giannessi (University of Pisa)
H.D. Sherali (Virginia Polytechnic and State University)
T. Terlaky (McMaster University)
Y. Ye (Stanford University)

Aims and Scope

Optimization has been expanding in all directions at an astonishing rate during the last few decades. New algorithmic and theoretical techniques have been developed, the diffusion into other disciplines has proceeded at a rapid pace, and our knowledge of all aspects of the field has grown even more profound. At the same time, one of the most striking trends in optimization is the constantly increasing emphasis on the interdisciplinary nature of the field. Optimization has been a basic tool in all areas of applied mathematics, engineering, medicine, economics, and other sciences.

The *Springer in Optimization and Its Applications* series publishes undergraduate and graduate textbooks, monographs, and state-of-the-art expository work that focus on algorithms for solving optimization problems and also study applications involving such problems. Some of the topics covered include nonlinear optimization (convex and nonconvex), network flow problems, stochastic optimization, optimal control, discrete optimization, multi-objective programming, description of software packages, approximation techniques and heuristic approaches.

For other titles published in this series, go to
http://www.springer.com/series/7393

INTRODUCTION TO NONLINEAR AND GLOBAL OPTIMIZATION

By

Eligius M.T. Hendrix
Málaga University, Spain

Boglárka G.-Tóth
Budapest University of Technology and Economics, Hungary

 Springer

Eligius M.T. Hendrix
Department of Computer Architecture
Málaga University
Málaga, Spain
Eligius.Hendrix@wur.nl

Boglárka G.-Tóth
Department of Differential Equations
Budapest University of Technology
and Economics
Budapest, Hungary
bog@math.bme.hu

ISSN 1931-6828
ISBN 978-0-387-88669-5   ISBN 978-0-387-88670-1 (eBook)
DOI 10.1007/978-0-387-88670-1
Springer New York Dordrecht Heidelberg London

Library of Congress Control Number: 2010925226

Mathematics Subject Classification (2010): 49-XX, 90-XX, 90C26

Printed on acid-free paper

Springer is part of Springer Science+Business Media (www.springer.com)

Contents

Preface

This book provides a solid introduction for anyone who wants to study the ideas, concepts, and algorithms behind nonlinear and global optimization. In our experience instructing the topic, we have encountered applications of optimization methods based on easily accessible Internet software. In our classes, we find that students are more often scanning the Internet for information on concepts and methodologies and therefore a good understanding of the concepts and keywords is already necessary.

Many good books exist for teaching optimization that focus on theoretical properties and guidance in proving mathematical relations. The current text adds illustrations and simple examples and exercises, enhancing the reader's understanding of concepts. In fact, to enrich our didactical methods, this book contains approximately 40 algorithms that are illustrated by 80 examples and 95 figures. Additional comprehension and study is encouraged with numerous exercises. Furthermore, rather than providing rigorous mathematical proofs, we hope to evoke a critical approach toward the use of optimization algorithms. As an alternative to focusing on the background ideas often furnished on the Internet, we would like students to study pure pseudocode from a critical and systematic perspective.

Interesting models from an optimization perspective come from biology, engineering, finance, chemistry, economics, etc. Modeling optimization problems depends largely on the discipline and on the mathematical modeling courses that can be found in many curricula. In Chapter 2 we use several cases from our own experience and try to accustom the student to using intuition on questions of multimodality such as "is it natural that a problem has several local optima?" Examples are given and exercises follow. No formal methodology is presented other than using intuition and analytic skills.

In our experience, we have observed the application of optimization methods with an enormous trust in clicking buttons and accepting outcomes. It is often thought that what comes out of a computer program must be true. To have any practical value, the outcomes should at least fulfill optimality conditions. Therefore in Chapter 3, we focus on the criteria of optimality illustrated with simple examples and referring further to the earlier mentioned books that present more mathematical rigor. Again, many exercises are provided.

The application and investigation of methods, with a nearly religious belief in concepts, like evolutionary programming and difference of convex programming, inspired us to explain such concepts briefly and then ask questions on the effectiveness and efficiency of these methods. Specifically, in Chapter 4 we pose questions and try to show how to investigate them in a systematic way. The style set in this chapter is then followed in subsequent chapters, where multiple algorithms are introduced and illustrated.

Books on nonlinear optimization often describe algorithms in a more or less explicit way discussing the ideas and their background. In Chapter 5, a uniform way of describing the algorithms can be found and each algorithm is illustrated with a simple numerical example. Methods cover one-dimensional optimization, derivative-free optimization, and methods for constrained and unconstrained optimization.

The ambition of global optimization algorithms is to find a global optimum point. Heuristic methods, as well as deterministic stochastic methods, often do not require or use specific characteristics of a problem to be solved. An interpretation of the so-called "no free lunch theorem" is that general-purpose methods habitually perform worse than dedicated algorithms that exploit the specific structure of the problem. Besides using heuristic methods, deterministic methods can be designed that give a guarantee to approach the optimum to an accuracy if structure information is available and used.

Many concepts exist which are popular in mathematical research on the structures of problems. For each structure at least one book exists and it was a challenge for us to describe these structures in a concise way. Chapter 6 explores deterministic global optimization algorithms. Each concept is introduced and illustrated with an example. Emphasis is also placed on how one can recognize structure when studying an optimization problem. The approach of branch and bound follows which aims to guarantee reaching a global solution while using the structure. Another approach that uses structure, the generation of cuts, is also illustrated. The main characteristic of deterministic methods is that no (pseudo-)random variable is used to find sample points. We start the chapter discussing heuristics that have this property. The main idea there is that function evaluations may be expensive. That means that it may require seconds, minutes, or even hours to find the objective function value of a suggested sample point.

Stochastic methods are extremely popular from an application perspective, as implementations of algorithms can be found easily. Although stochastic methods have been investigated thoroughly in the field of global optimization, one can observe a blind use of evolution-based concepts. Chapter 7 tries to summarize several concepts and to describe algorithms as basically and as dryly as possible, each illustrated. Focus is on a critical approach toward the results that can be obtained using algorithms by applying them to optimization problems.

We thank all the people who contributed to, commented on, and stimulated this work. The material was used and tested in master's and Ph.D. courses at the University of Almería where colleagues were very helpful in reading and commenting on the material. We thank the colleagues of the Computer Architecture Department and specifically its former director for helping to enhance the appearance of the book. Students and colleagues from Wageningen University and from Budapest University of Technology and Economics added useful commentary. Since 2008 the Spanish ministry of science has helped by funding a Ramón y Cajal contract at the Computer Architecture department of Málaga University. This enabled us to devote a lot of extra time to the book. The editorial department of Springer helped to shape the book and provided useful comments of the anonymous referees.

Eligius M.T. Hendrix
Boglárka G.-Tóth
September 2009

1

Introduction

1.1 Optimization view on mathematical models

Optimization can be applied to existing or specifically constructed mathematical models. The idea is that one would like to find an extreme of one output of the model by varying several parameters or variables. The usual reason to find appropriate parameter values is due to decision support or design optimization. In this work we mainly consider the mathematical model as given and have a look at how to deal with optimization. Several examples of practical optimization problems are given.

The main terminology in optimization is as follows. Usually quantities describing the decisions are given by a vector $x \in \mathbb{R}^n$. The property (output) of the model that is optimized (costs, CO_2 emission, etc.) is put in a so-called objective function $f(x)$. Other relevant output properties are indicated by functions $g_i(x)$ and are put in constraints representing design restrictions such as material stress, $g_i(x) \leq 0$ or $g_i(x) = 0$. The so-called feasible area that is determined by the constraints is often summarized by $x \in X$. In nonlinear optimization, or nonlinear programming (NLP), the objective and/or constraint functions are nonlinear.

Without loss of generality the general NLP problem can be written as

$$\min f(x) \text{ subject to}$$
$$g_i(x) \leq 0 \text{ for some properties } i, \text{ inequality constraints,} \qquad (1.1)$$
$$g_i(x) = 0 \text{ for some properties } i, \text{ equality constraints.}$$

The general principle of NLP is that the values of the variables can be varied in a continuous way within the feasible set. To find and to characterize the best plan (suggestion for the values of the decision variables), we should define what is an optimum, i.e., maximum or minimum. We distinguish between a local and global optimum, as illustrated in Figure 1.1. In words: a plan is called locally optimal, when the plan is the best in its neighborhood. The plan is called globally optimal, when there is no better plan in the rest of the feasible area.

E.M.T. Hendrix and B.G.-Tóth, *Introduction to Nonlinear and Global Optimization*, Springer Optimization and Its Applications 37, DOI 10.1007/978-0-387-88670-1_1, © Springer Science+Business Media, LLC 2010

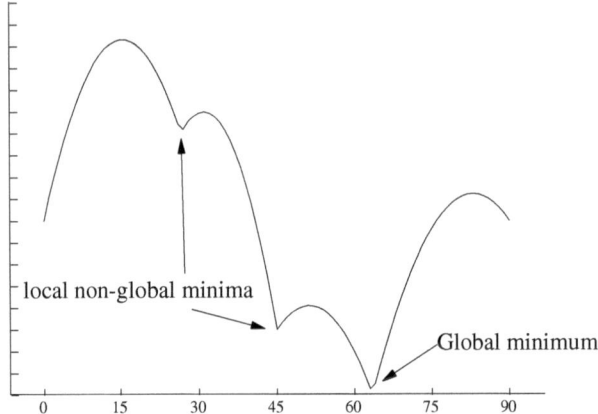

local non-global minima

Global minimum

Fig. 1.1. Global optimum and local optima

In general one would try to find an optimum with the aid of some software, which is called an implementation of an algorithm. An algorithm is understood here to be a list of rules or commands to be followed by the calculation process in the computer. To interpret the result of the computer calculation (output of the software), the user should have some feeling about optimality criteria; is the result really an optimum and how can it be interpreted? In this book we distinguish between three important aspects which as such are of importance for different interest groups with respect to NLP.

- How to recognize an optimal plan? A plan is optimal when it fulfills so-called optimality conditions. Understanding of these conditions is useful for a translation to the practical decision situation. Therefore it is necessary to go into mathematical analysis of the underlying model. The topic of optimality conditions is explained in Chapter 3. This is specifically of interest to people applying the methods (software).
- Algorithms can be divided into NLP local search algorithms that given a starting point try to find a local optimum, and global optimization algorithms that try to find a global optimum, often using local optimizers multiple times. In Chapters 5, 6 and 7 we describe the ideas behind the algorithms with many numerical examples. These chapters are of interest to people who want to know how the underlying mechanisms work and possibly want to make implementations themselves.
- *Effectiveness* of algorithms is defined by their ability to reach the target of the user. *Efficiency* is the effort it costs to reach the target. Traditionally *mathematical programming* is the field of science that studies the behavior of optimization algorithms with respect to those criteria depending on the structure of the underlying optimization problem. Chapter 4 deals with the question of investigating optimization algorithms in a systematic way.

This is specifically of interest to researchers in Operations Research or mathematical programming.

The notation used throughout the book will stay close to using f for the objective function and x for the decision variables. There is no distinction between a vector or scalar value for x. As much as possible, index j is used to describe the component x_j and index k is used for iterates x_k in the algorithmic description. Moreover, we follow the convention of using boldface letters to represent stochastic variables. The remainder of this chapter is devoted to outlining the concept of formulating NLP problems.

1.2 NLP models, black-box versus explicit expression

Optimization models can be constructed directly for decision support following an Operations Research approach, or can be derived from practical numerical models. It goes too far to go into the art of modeling here. From the point of view of nonlinear optimization, the distinction that is made is due to what can be analyzed in the model. We distinguish between:

- Analytical expressions of objective function f and constraint functions g_i are available.
- The so-called black-box or oracle case, where the value of the functions can only be obtained by giving the parameter values (values for the decision variables x) to a subroutine or program that generates the values for f and/or g_i after some time.

This distinction is relevant with respect to the ability to analyze the problem, to use the structure of the underlying problem in the optimization and to analyze the behavior of algorithms for the particular problem.

Let us go from abstraction to a simple example, the practical problem of determining the best groundwater level. Let us assume that the groundwater level x should be determined, such that one objective function f is optimized. In practice this is not a simple problem due to conflicting interests of stakeholders. Now the highly imaginary case is that an engineer would be able to compose one objective function and write an explicit formula:

$$f(x) = 2x - 100 \ln(x), \; 30 \leq x \leq 70. \tag{1.2}$$

The explicit expression (1.2) can be analyzed; one can make a graph easily and calculate the function value for many values of the variable x in a short time. However, often mechanistic models are used that describe the development of groundwater flows from one area to the other. The evaluation of an (or several) objective function may take minutes, hours or days on a computer when more and more complicated and extended descriptions of flows are included. In the optimization literature the term *expensive function evaluations* is also used. This mainly refers to the question that the evaluation of the model is

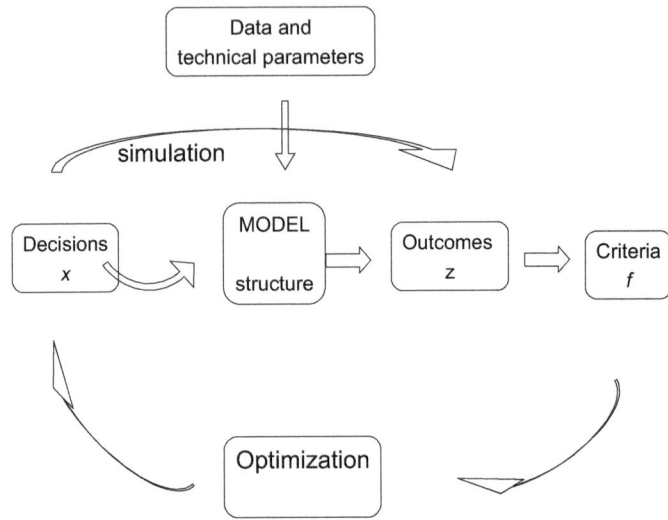

Fig. 1.2. Optimization in mathematical models

relatively time consuming compared the algorithmic operations to generate new trial points.

From the optimization point of view, the terms oracle or black-box case are used, as no explicit expression of the objective function is visible. From the modeler point of view, this is the other way around, as the mechanistic model is closer to the processes that can be observed in reality and expression (1.2) does not necessarily have any relation with the physical processes. Figure 1.2 sketches the idea. In general a mathematical model has inputs (all kinds of technical parameters and data) and outputs. It becomes an optimization problem for decision support as soon as performance criteria are defined and it has been assessed as to what are the input parameters that are considered to be variable. Running a model or experimenting with it, is usually called simulation. From the point of view of optimization, giving parameter values and calculating criteria f is called a *function evaluation*. Algorithms that aim at finding "a" or "the" optimum, usually evaluate the function many times. The efficiency is measured by the number of function evaluations related to the calculation time that it requires per iteration. Algorithms are more specific when they make more use of the underlying *structure* of the optimization problem. One of the most successful and applied models in Operations Research is that of Linear Programming, where the underlying input–output relation is linear.

One type of optimization problems concerns parameter estimation problems. In statistics when the regression functions are relatively simple the term *nonlinear regression* is used. When we are dealing with more complicated models (for instance, using differential equations), the term *model calibration* is used. In all cases one tries to find parameter values such that the output of the model fits well observed data of the output according to a certain fitness criterion. In the next chapter a separate section is devoted to sketching problems of this type.

In Chapter 2, the idea of black-box modeling and explicit expressions is illustrated by several examples from the experience of working 20 years with engineering and economic applications. Exercises are provided to practice with the idea of formulating nonlinear programming models and to study whether they may have more than one local optimum solution.

2

Mathematical modeling, cases

2.1 Introduction

This chapter focuses on the modeling of optimization problems where objective
and constraint functions are typically nonlinear. Several examples of practical
optimization cases based on our own experience in teaching, research and
consultancy are given in this chapter. The reader can practice by trying
to formulate the exercise examples based on the cases at the end of the
chapter.

2.2 Enclosing a set of points

One application that can be found in data analysis and parameter identification is to enclose a set of points with a predefined shape with a size or
volume as small as possible. Depending on the enclosure one is looking for, it
can be an easy to solve problem, or very hard. The first problem is defined as:

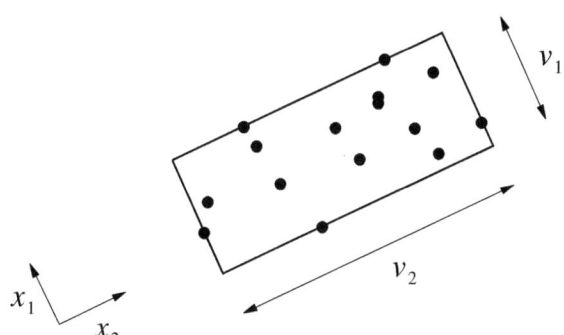

Fig. 2.1. Minimum volume hyperrectangle problem

E.M.T. Hendrix and B.G.-Tóth, *Introduction to Nonlinear and Global Optimization*,
Springer Optimization and Its Applications 37, DOI 10.1007/978-0-387-88670-1_2,
© Springer Science+Business Media, LLC 2010

given the set of points $P = \{p_1, \ldots, p_K\} \in \mathbb{R}^n$, find an enclosing hyperrectangle with minimum volume around the points of which the axes are free to be chosen; see Keesman (1992). Mathematically this can be translated into finding an orthonormal matrix $X = (x_1, \ldots, x_n)$ minimizing the objective

$$f(X) = \prod_{i=1}^{n} \nu_i, \tag{2.1}$$

where $\nu_i = (\max_{j=1,\ldots,K} x_i^T p_j - \min_{j=1,\ldots,K} x_i^T p_j)$, the length of the edges of the hyperrectangle as in Figure 2.1. Here the axes x are seen as decision variables and the final objective function f consists of a multiplication of the lengths ν_i that appear after checking all points. So the abstract model of Figure 1.2 can be filled in as checking all points p_j over their product with x_i. In Figure 2.2 the set $P = \{(2,3), (4,4), (4,2), (6,2)\}$ is enclosed by rectangles defined by the angle α of the first axis x_1, such that vector $x_1 = (\cos\alpha, \sin\alpha)$. This small problem has already many optima, which is illustrated by Figure 2.3.

Note that case $\alpha = 0$ represents the same situation as $\alpha = 90$, because the position of the two axes switches. The general problem is not easy to formulate explicitly due to the orthonormality requirements. The requirement of the orthonormality of the matrix of axes of the hyperrectangle, implies the degree of freedom in choosing the matrix to be $n(n-1)/2$. In two dimensions this can be illustrated by using one parameter, i.e., the angle of the first vector. In higher dimensions this is not so easy. The number of optima as such is enormous, because it depends on the number of points, or more precisely, on the number of points in the convex hull of P.

A problem similar to the minimum volume hyperrectangle problem is to find an enclosing or an inscribed ellipsoid, as discussed for example by Khachiyan and Todd (1993). The enclosing minimum volume ellipsoid

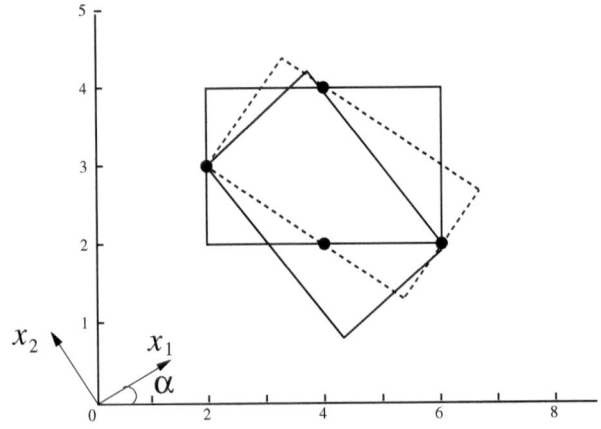

Fig. 2.2. Rectangles around four points

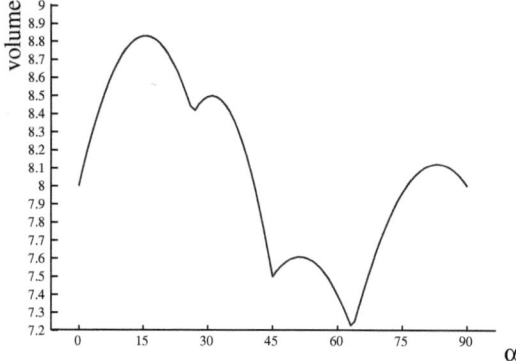

Fig. 2.3. Objective value as a function of angle

problem can be formulated as finding a positive definite matrix and a center of the ellipsoid, such that it contains a given set of points or a polytope. This problem is fairly well analyzed in the literature, but goes too far to be formulated as an example.

Instead, we focus on the so-called *Chebychev (centroid location)* problem of finding the smallest sphere or ball around a given point set. For lower dimensions, the interpretation in locational analysis is to locate a facility that can reach all demand points as fast as possible. Given a set of K points $\{p_1, \ldots, p_K\} \in \mathbb{R}^n$, find the center c and radius r such that the maximum distance over the K points to center c is at its minimum value. This means, find a sphere around the set of points with a radius as small as possible. In $\mathbb{R}^2$ this problem is not very hard to solve. In Figure 2.4, an enclosing sphere is given that does not have the minimum radius. In general, the optimal center c is called the Chebychev center of a set of points or the 1-center of demand points.

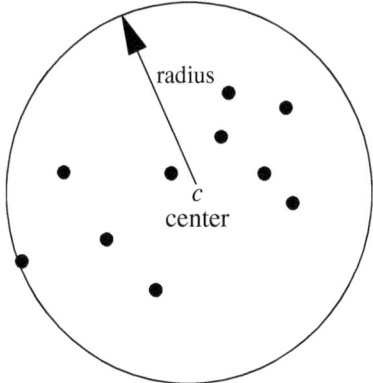

Fig. 2.4. Set of points with an enclosing sphere

2.3 Dynamic decision strategies

The problems in this section involve sequential decision making. The performance, objective function, not only depends on the sequence of decisions, but also on fluctuating data over a given time period, often considered as a stochastic variable. The calculation of the objective typically requires the simulation of the behavior of a system over a long period.

The first example is derived from an engineering consulting experience dealing with operating rules for pumping water into a higher situated lake in the Netherlands. In general the rainfall exceeds the evaporation and the seepage. In summer, however, water has to be pumped from lower areas and is treated to maintain a water level above the minimum with a good water quality. Not only the pumping, but certainly also the treatment to remove phosphate, costs money. The treatment installation performs better when the stream is constant, so the pumps should not be switched off and on too frequently. The behavior of the system is given by the equation

$$I_t = \min\{I_{t-1} + \xi_t + x_t, \text{Max}\} \tag{2.2}$$

with

I_t: water level of the lake
ξ_t: natural inflow, i.e., rainfall - seepage - evaporation
x_t: amount of water pumped into the lake
Max: maximum water level.

Figure 2.5 depicts the situation. When the water level reaches its maximum (Max), the superfluous water streams downwards through a canal system toward the sea. For the studied case, two pumps were installed, so that x_t only takes values in $\{0, B, 2B\}$, where B is the capacity of one pump. Decisions are taken on a daily basis. In water management, it is common practice

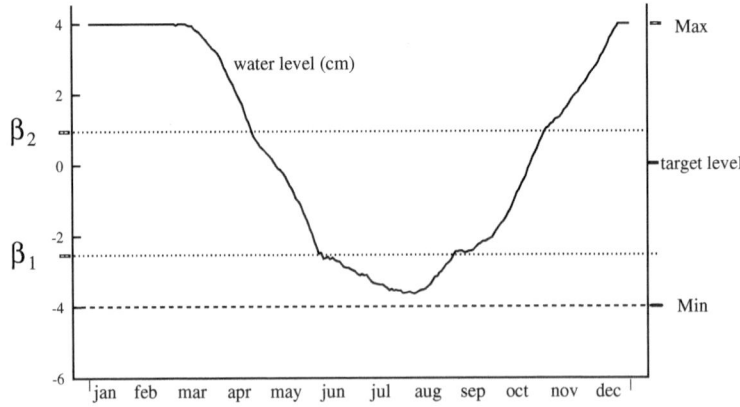

Fig. 2.5. Strategy to rule the pumping

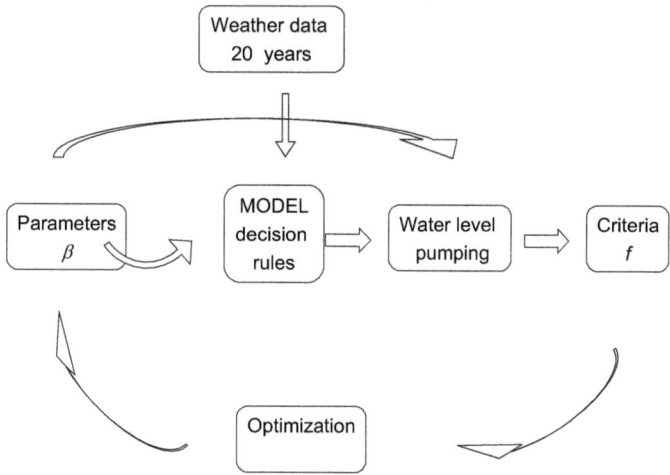

Fig. 2.6. Determining parameter values

to derive so-called operating rules, decision strategies including parameters. A decision rule instructs as to what decision to make in which situation. An example is rule (2.3) with parameters β_1 and β_2:

$$
\begin{array}{ll}
I_t < \beta_1 & x_t = 2B \\
\beta_1 \le I_t \le \beta_2 & x_t = B \\
I_t > \beta_2 & x_t = 0.
\end{array}
\tag{2.3}
$$

Given weather data of a certain period, now the resulting behavior of a sequence of decisions x_t can be evaluated by measuring performance indicators such as the amount of water pumped $\sum x_t$ and the number of switches of the pumps $\sum | x_t - x_{t-1} | / B$. Assessment of appropriate values for the parameters β_1 and β_2 can be considered as a black-box optimization problem.

The total idea is captured in Figure 2.6. For every parameter set, the model (2.2) with strategy (2.3) can be simulated with weather data (rainfall and evaporation) of a certain time period. Some 20 years of data on open water evaporation and rainfall were available. The performance can be measured leading to one (multi)objective function value. At every iteration an optimization algorithm delivers a proposal for the parameter vector β, the model simulates the performance and after some time returns an objective function value $f(\beta)$. One possibility is to create a stochastic model of the weather data and resulting ξ_t and to use the model to "generate" more years by Monte Carlo simulation, i.e., simulation using (pseudo) random numbers. In this way, it is possible to extend the simulation run over many years. The model run can be made arbitrarily long. Notice that in our context it is useful for every parameter proposal to use the same set of random numbers (seed), otherwise the objective function $f(\beta)$ becomes a random variable. The problem sketched here is an example of so-called *parametrized decision strategies*.

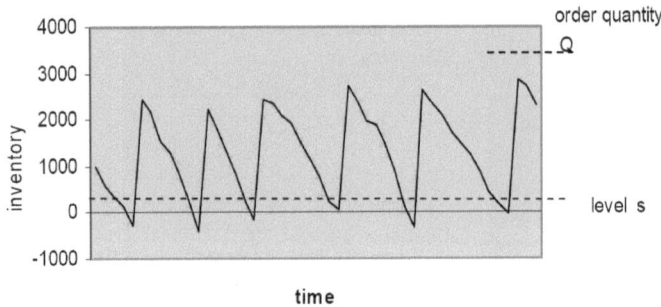

Fig. 2.7. Inventory level given parameter values

An important application in Logistics is due to Stochastic Inventory Control. Equation (2.2) now reads as follows (see Hax and Candea (1984)):

I_t: level of inventory
x_t: amount produced or ordered
ξ_t: (negative) demand, considered stochastic.

Let us consider a so-called (s, Q)-policy. As soon as the inventory is below level s, an order is placed of size Q. An order becomes available at the end of the next day. If there is not sufficient stock (inventory), the client is supplied the next day (back ordering) at additional cost. Criteria that play a role are inventory holding cost, ordering cost and back ordering or out of stock cost.

Usually in this type of problem an analysis is performed, based on integrating over the probability density function of the uncertain demand and delivery time. We will consider the problem based on another approach. As a numerical example, let inventory holding cost be 0.3 per unit per day, ordering cost be 750 and back order cost be 3 per unit. Figure 2.7 gives the development of inventory following the (s, Q) system based on samples of a distribution with an average demand of 400. Let us for the exercise, consider what is the optimal order quantity Q if the demand is not uncertain, but fixed at 400 every day, the so-called *deterministic* situation. The usual approach is to minimize the average daily costs. The length of a cycle in the sawtooth figure is defined by $Q/400$, such that the order cost per day is $750/(Q/400) = 300000/Q$. The average inventory costs are derived from the observation that the average inventory over one cycle is $Q/2$. The total relevant cost per day, $TRC(Q)$, is given by

$TRC(Q) = 300000/Q + 0.3 \cdot Q/2.$

- What is the order quantity Q that minimizes the total daily cost for this deterministic situation?

A second way to deal with the stochastic nature of the problem is to apply Monte Carlo simulation. Usually this is done for more complex studies, but a small example can help us to observe some generic problems that appear.

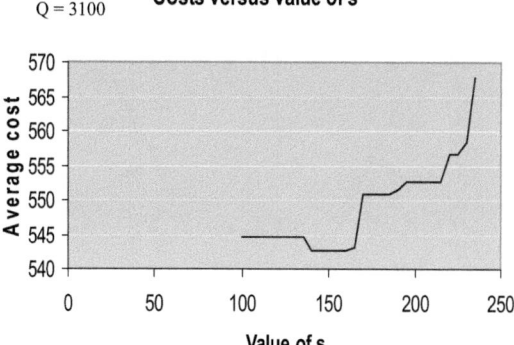

Fig. 2.8. Average cost as a function of value for s applying a finite number of scenarios (Monte Carlo)

Assume we generate 100 data for the demand, or alternatively use the data of 100 days. Fixing the value $Q = 3100$ and varying the value of s gives the response in Figure 2.8. The discontinuities and local insensitivities appear due to IF-THEN constructions in the modeling. This makes models of this type hard to optimize, see Hendrix and Olieman (2008).

2.4 A black box design; a sugar centrifugal screen

An example is sketched of a design problem where from the optimization point of view, the underlying model is a black-box (oracle) case. The origin of this case is due to a project in cooperation with a metallurgic firm which among others produces screens for sugar refiners. The design parameters, of which several are sketched in Figure 2.9, give the degree of freedom for the

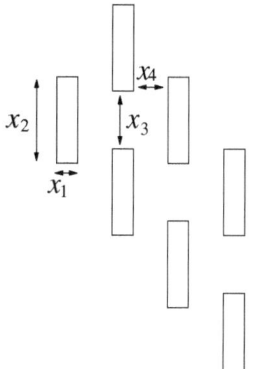

Fig. 2.9. Parameters of slot grid pattern

Fig. 2.10. Sugar refiner screen

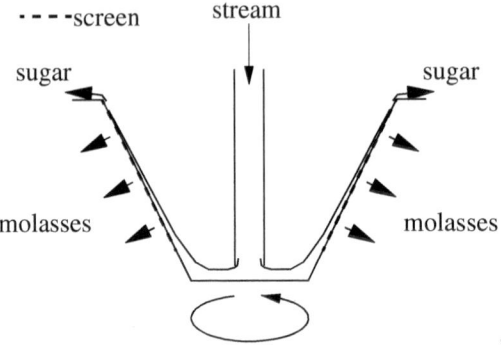

Fig. 2.11. Continuous centrifugal

product development group to influence the pattern in sugar screens. The quality of a design can be evaluated by a mathematical model. This describes the behavior of the filtering process where sugar is separated from molasses in a continuous sugar centrifugal. We first give a flavor of the mathematical model. The continuous centrifugal works as follows (Figure 2.11). The fluid (molasses) including the sugar crystals streams into the middle of the rotating basket. By the centrifugal force and the angle of the basket, the fluid streams uphill. The fluid goes through the slots in the screen whereas the crystals continue their way uphill losing all fluid which is still sticking on the material. Finally the crystals are caught at the top of the basket. The constructed model describes the stream of the fluid from the start, down in the basket, until the end, top of the screen. The flux of the fluid through the screen does not only depend on the geometry of the slots, but also on the centrifugal force and height of the fluid film on a certain position. Reversely, the height depends on how quickly the fluid goes through the screen. Without going into detail, this interrelation can be described by a set of differential equations which can be solved numerically. Other relations were found to describe the strength of the screen, as wear is a big problem.

In this way a model is sketched in the sense of Figure 1.2, which given technical data such as the size and angle of the basket, revolutions per second, the stream into the refiner, the viscosity of the material, the shape of the slots and the slot grid pattern, calculates the behavior described by the fluid profile and the strength of the screen. Two criteria were formulated; one to describe the strength of the screen and one to measure the dryness of the resulting sugar crystals. There are several ways to combine the two criteria in a multicriteria approach. Actually we are looking for several designs on the so-called Pareto set describing screens which are strong and deliver dry sugar crystals when used in the refiner. Several designs were generated that were predicted to perform better than existing screens. The use of a mathematical model in this design context is very useful, because it is extremely difficult to do real life experiments. The approach followed here led to an advisory system to make statements on what screens to use in which situation. Furthermore, it led to insights for the design department which generated and tested several new designs for the screens.

2.5 Design and factorial or quadratic regression

Regression analysis is a technique which is very popular in scientific research and in design. Very often it is a starting point for the identification of relations between inputs and outputs of a system. In a first attempt one tries to verify a linear relation between output y, called regressand or dependent variable, and the input vector x, called regressor, factor or independent variable. A so-called linear regression function is used:

$$y = \beta_0 + \beta_1 x_1 + \beta_2 x_2 + \cdots + \beta_n x_n.$$

For the estimation of the coefficients β_j and to check how good the function "fits reality," either data from the past can be used or experiments can be designed to create new data for the output and input variables. The data for the regression can be based on a design of a computer experiment which uses a simulation model to generate the data on input and output. The generation of regression relations out of experiments of a relatively large simulation model is called metamodeling and is discussed in Kleijnen and van Groenendaal (1988). The regression model is called a metamodel, because it models the input–output behavior of the underlying simulation model. In theory about design, the term response surface methodology is more popular and promoted by Taguchi among others; see Taguchi et al. (1989) and Box and Draper (2007).

The regression functions based on either historical data, special field experiments or computer experiments can be used in an optimization context. As long as the regression function is linear in the parameters β, and in the input variables x_j, linear programming can be applied. The optimization becomes more complicated when interaction between the input variables is introduced in the regression function. Interaction means that the effect of an input variable depends on the values of another input variable. This is usually introduced by allowing so-called two-factor interaction, i.e., multiplications of two input variables in the regression function. An example of such a factorial regression model is

$$y = \beta_0 + \beta_1 x_1 + \beta_2 x_2 + \beta_{12} x_1 x_2. \tag{2.4}$$

The introduction of multiplications implies the possibility to have several optima in an optimization context.

Example 2.1. Consider the minimization of $y(x) = 2 - 2x_1 - x_2 + x_1 x_2$ with $0 \le x_1 \le 4$ and $0 \le x_2 \le 3$. This problem has two minima: $y = -1$ for $x = (0, 3)$ and $y = -6$ for $x = (4, 0)$.

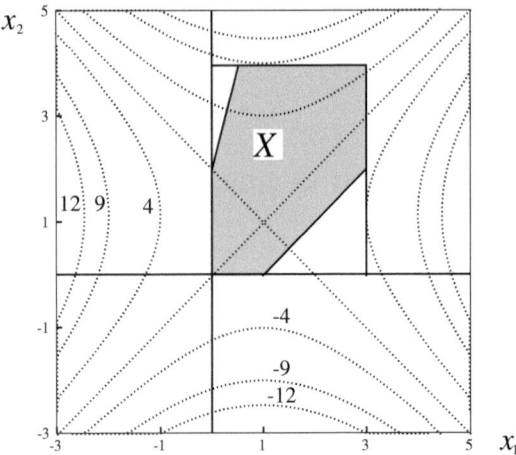

Fig. 2.12. Indefinite quadratic problem

A further extension in regression analysis is to complete the second-order Taylor series approximation, which is called quadratic regression to distinguish from (2.4). In two dimensions the quadratic regression function is

$$y = \beta_0 + \beta_1 x_1 + \beta_2 x_2 + \beta_{12} x_1 x_2 + \beta_{11} x_1^1 + \beta_{22} x_2^2.$$

Example 2.2. Consider the following Indefinite Quadratic Program:

$$\min_{x \in X}\{f(x) = (x_1 - 1)^2 - (x_2 - 1)^2\}$$
where X is given by
$$x_1 - x_2 \leq 1$$
$$4x_1 - x_2 \geq -2$$
$$0 \leq x_1 \leq 3, \quad 0 \leq x_2 \leq 4.$$

Contour lines and the feasible set are given in Figure 2.12. The problem has two local minimum points, i.e., $(1, 0)$ and $(1, 4)$ (the global one).

Notice that in regression terms this is called linear regression, as the function is linear in the parameters β. When these functions are used in an optimization context, it depends on the second-order derivatives β_{ij} whether the function is convex and consequently whether it may have only one or multiple optima such as in Example 2.2.

The use in a design case is illustrated here with the mixture design problem, which can be found in Hendrix and Pintér (1991) and in Box and Draper (2007). An illustration is given by the so-called rum–coke example.

Example 2.3. A bartender tries to find a mix of rum, coke, and ice cubes, such that the properties $y_i(x)$ fulfill the following requirements:

$$y_1(x) = -2 + 8x_1 + 8x_2 - 32x_1x_2 \leq -1$$
$$y_2(x) = 4 - 12x_1 - 4x_3 + 4x_1x_3 + 10x_1^2 + 2x_3^2 \leq 0.4.$$

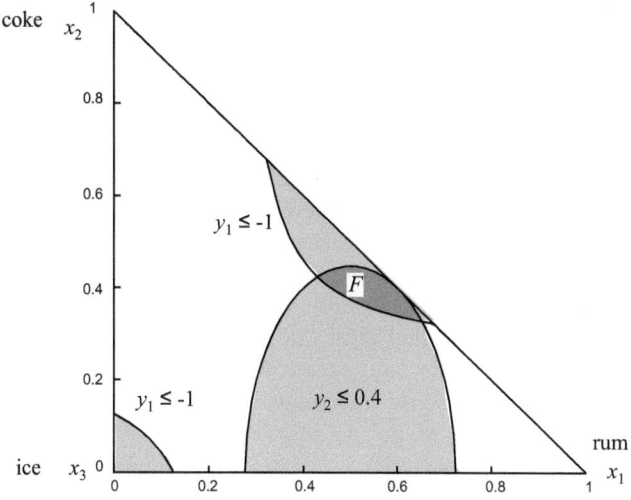

Fig. 2.13. Rum–coke design problem

The area in which the mixture design problem is defined is given by the unit simplex S, where $x_1 + x_2 + x_3 = 1$. Projection of the three-dimensional unit simplex S on the x_1, x_2 plane gives the triangle as in Figure 2.13. Vertex x_p represents a product consisting of 100% of component p, $p = 1, 2, 3$ (rum, coke and ice cubes). The area in which the feasible products are situated is given by F. One could try to find a feasible design for a design problem defined by inequalities $y_i(x) \le b_i$, by minimizing an objective function

$$f(x) = \max_i \{y_i(x) - b_i\} \tag{2.5}$$

or by minimizing

$$f(x) = \sum_i \max\{y_i(x) - b_i, 0\}. \tag{2.6}$$

The problem of minimizing (2.6) over S has a local optimum in $x_{loc} = (0.125, 0, 0.875)$, $f(x_{loc}) = -0.725$, and of course a global optimum $(= 0)$ for all elements of $F \cap S$.

2.6 Nonlinear optimization in economic models

Economics studies human behavior in its relation with scarce resources. The concept that economic agents (homo economicus) act in a rational optimizing way makes the application of optimization popular. In Chapter 3, some small examples are given derived from ideas in micro-economics. Many times, however, models with many decision variables are used to describe the behavior of an economic system. The multiplicity of variables is caused by defining separate variables for various aspects:

- Multiple agents, consumers, producers, countries, farmers, etc.
- Distinguishing spatial units, regions, plots, for which one takes decisions.
- Temporal elements (time), years, months, weeks, etc., are used.

The underlying mathematical structure of the models enhances the ideas of decreasing returns to scale and diminishing marginal utility. These tendencies usually cause the model to have one optimum. The difficulty is the dimensionality; so many aspects can be added such that the number of variables explodes and cannot be handled by standard software.

Usually so-called modeling languages are used to formulate a model with hundreds of variables. After the formulation the model is fed to a so-called solver, an implementation of an optimization algorithm, and a solution is fed back to the modeling software. The GAMS-software (www.gams.com) is used frequently in this field. However, there are also other systems available such as AMPL (www.ampl.com), Lingo (www.Lindo.com) and AIMMS (www.aimms.com). In the appendix, a small example is shown in the GAMS format. To give a flavor of the type of models, examples from Wageningen University follow.

2.6.1 Spatial economic-ecological model

In economic models where an economic agent is taking decisions on spatial entities, decision variables are distinguished for every region, plot, grid cell, etc. As an illustration we take some elements from Groeneveld and van Ierland (2001), who describe a model where a farmer is deciding on the use of several plots for cattle feeding and the consequence on biodiversity conservation. Giving values to land-use types l for every plot p for every season t, determines in the end the costs needed for fodder (economic criterion) and the expected population size of target species (ecological criterion) with a minimum population size s. The latter criterion is in fact a nonlinear function of the land use, but in their paper is described by piecewise linear functions. The restrictions have a typical form:

$$\sum_{p} K_{pt} = \kappa \qquad \forall t \tag{2.7}$$

$$\left(\sum_{l} F_{plt}\alpha_p q_{plt}\right) + B_{pt}^{growing\ season} \geq \phi K_{pt} \qquad \forall p,t. \tag{2.8}$$

In their model, variable F_{plt} denotes the fraction of land-use type l on plot p in season t, variable K_{pt} denotes the number of animals per plot p and B_{pt} denotes the amount of fodder purchased. Without going into detail of

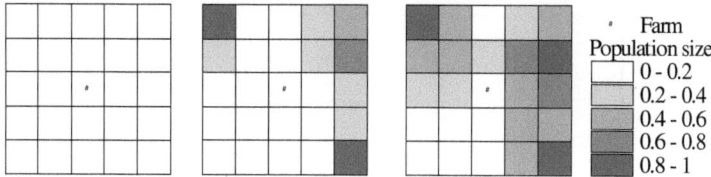

Fig. 2.14. Outcome example of the spatial model; population of the target species in numbers per plot for $s = 0$ (left), $s = 5$ (center) and $s = 10$ (right)

used values for the parameters, which can be found in the article, what we learn from such models is that summation symbols, $\sum$, are used and many variables are defined. Equations like (2.7) and (2.8) can directly be translated into modeling languages. Figure 2.14 illustrates how outcomes of a spatial model can be shown in a graphical way.

2.6.2 Neoclassical dynamic investment model for cattle ranching

In a dynamic model, the decision variables have a time aspect. In continuous optimization, one speaks of *optimal control* when the decision sequence is considered with infinitely many small time steps and the outcome is a continuous trajectory. Often in economic models the time is considered with discrete periods (year, month, week, etc.). In this case a model can be formulated in a nonlinear optimization way. In Roebeling (2003) the traditional neoclassical investment model is reformulated for pasture cattle production in order to study effects of price of land.

$$\text{Maximize} \quad \sum_t (1 + r)^{-t}[pQ(S_t, A_t) - p_S S_t - p_A A_t - c(I_t)] \qquad (2.9)$$

$$\text{subject to} \quad \begin{aligned} A_t &= A_{t-1} + I_t \quad t = 1, \ldots \text{ (equation of motion for } A_t) \\ A_0 &> 0 \text{ and } I_0 = 0 \qquad \text{(initial conditions)} \\ A_t &\geq 0 \text{ and } S_t \geq 0. \end{aligned}$$

The decision variables of the model include decisions on cattle stock S, investment in land I and a resulting amount of pasture area A for every year t. As in an economic model typically we have functions to describe the production Q and costs c. The dynamic structure is characterized by an equation of motion that describes the dynamics, the relation between the time periods. The concept of discounting is used in the model (2.9). The final optimal path for the decision variables depends on the time horizon and the prices of end-product (p), maintenance of land (p_A) and cattle (p_S) and interest rate r. A typical path is given in Figure 2.15.

2.6.3 Several optima in environmental economics

The following model describes a simple example of one pollutant having several abatement techniques to reduce its emission:

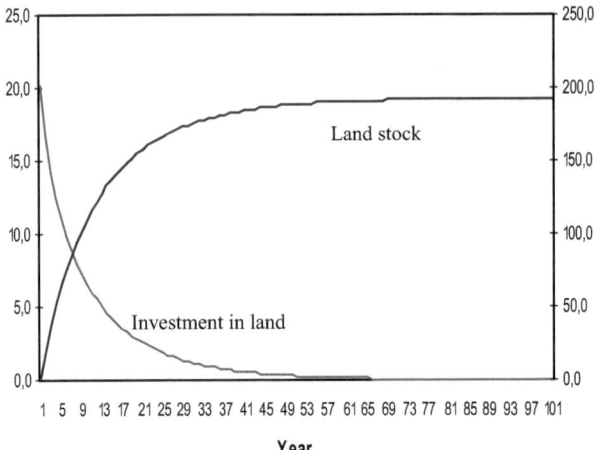

Fig. 2.15. Typical outcome of a dynamic model

$$\min \sum_i C_i(x_i)$$
$$\text{subject to } \; Em \cdot R \le \epsilon$$

with

$$C_i(x_i) = \alpha_i \cdot x_i$$
$$R = \prod_i (1 - x_i \cdot \rho_i)$$
$$0 \le x_i \le 1$$

(2.10)

x_i: implementation rate abatement technique i
R: fraction of remaining emissions
Em: emissions before abatement
$C_i(x)$: cost function for abatement technique i
ϵ: emission target
α_i: cost coefficient technique i
ρ_i: fraction emissions reduced by technique i.

The resulting iso-cost and iso-emission lines for two reduction techniques are depicted in Figure 2.16. Typical in the formulation here is that looking for minimum costs would lead to several optimum solutions. Due to the structure of decreasing return to scales and decreasing marginal utility, this rarely happens in economic models. In the example here, it is caused by the multiplicative character of the abatement effects. A similar problem with more pollutants, more complicated abatement costs, many abatement techniques can easily be implemented into a modeling language. The appearance of several optima will persist.

2.7 Parameter estimation, model calibration, nonlinear regression

A problem solved often by nonlinear optimization is due to parameter estimation. In general a mathematical model is considered good when it describes

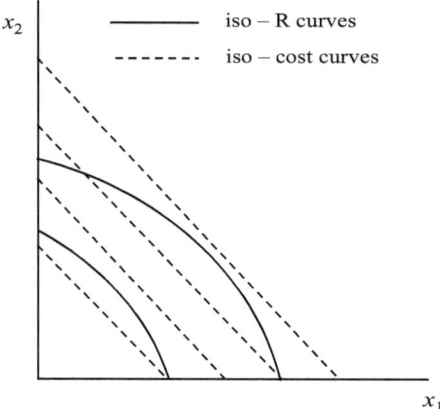

Fig. 2.16. Graphical representation for two reduction techniques

the image of the object system in the head of the modeler well; it "fits reality." Model validation is—to put it in a simple and nonmathematical way—a matter of comparing the calculated, theoretical model results with measured values. One tries to find values for parameters, such that input and output data fit relatively well, as depicted in Figure 2.17. In statistics, when the regression models are relatively simple, the term *nonlinear regression* is used.

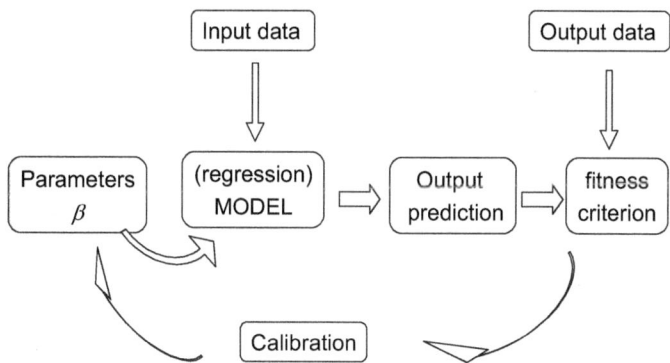

Fig. 2.17. Calibration as an optimization problem

When we are dealing with more complicated (for instance using differential equations) models, the term *model calibration* is used. In all cases one tries to find parameter values such that the output of the model fits well observed data of the output according to a certain fitness criterion.

For the linear regression type of models, mathematical expressions are known and relatively easy. Standard methods are available to determine

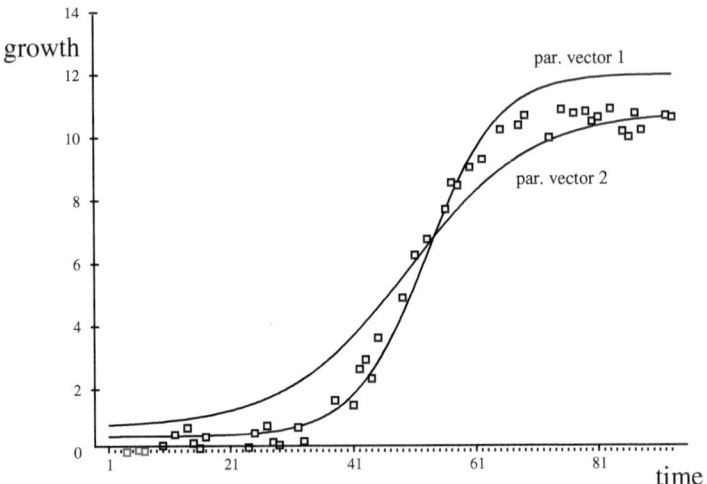

Fig. 2.18. Two growth curves confronted with data

optimal parameter values and an optimal experimental design. For the general *nonlinear regression* problem, this is more or less also the case. For instance, when using growth curves, which are popular in environmental sciences and biology, methods appear to be available to estimate parameters and to find the best way to design experiments; e.g., Rasch et al. (1997).

The mathematical expressions in the standard models have been analyzed and used for the derivation of algorithms. Formalizing the parameter estimation problem, the output variable y is explained by a model given input variable x and parameter vector β. In nonlinear regression, the model $z(x, \beta)$ is called a regression function, y is called the regressand and x is called the regressor. When measurements $i = 1, \ldots, m$ are available of regressand y_i and regressor x_i, the model calculations $z(x_i, \beta)$ can be confronted with the data y_i. The discrepancy $e_i(\beta) = z(x_i, \beta) - y_i$ is called the residual or error. Figure 2.18 illustrates this confrontation. Data points representing time x and growth y are depicted with two parameter value vectors for a logistic growth curve which is common in biological and environmental sciences,

$$z(x, \beta) = \frac{\beta_1}{1 + \beta_2^{\beta_3 x}}. \qquad (2.11)$$

In Figure 2.18, the curve of parameter vector 1 fits well the data at the beginning of the curve, whereas the curve of parameter vector 2 fits better the data at the end. The discrepancy measure or goodness of fit criterion, combines the residual terms in a multiobjective way. There are numerous ways of doing so. Usually one minimizes the sum of (weighted) absolute or squared values of the error terms:

$$f(\beta) = \sum |e_i(\beta)|, \quad \text{or} \qquad (2.12)$$

$$f(\beta) = \sum e_i^2(\beta). \tag{2.13}$$

A less frequently used criterion is to look at the maximum absolute error $\max_i | e_i(\beta) |$ over the observations. The minimization of squared errors (2.13) has an important interpretation in statistics. When assumptions are made such as that the measurement errors are independent normally distributed random variables, the estimation of β by minimizing $f(\beta)$ of (2.13) corresponds to a so-called maximum likelihood estimate and probabilistic statements can be made; see, e.g., Bates and Watts (1988). Parameter estimation by minimizing (2.13) given data on y_i and x_i is called an ordinary *least squares* approach.

In general more complicated models, which make use of sets of differential equations, are applied to describe complex systems. Often the structure is hidden from the optimization point of view. Although the structure of time series and differential equations can be used for the derivation of appropriate methods, this is complicated as several equations are involved and measurements in general concern several output variables of the model.

Interpretation of local optima
The function which is optimized in parameter estimation, values the discrepancy between model calculations and measurements. As the data concern various observations of possibly several output variables, the function is inherently a multiobjective function; it has to combine the discrepancies of all observations. A locally optimal parametrization (values for the parameters) can indicate that the model fits well one part of the observations whereas it describes other observations, or another output variable, badly. One optimum can give a good description of downstream measurements in a river model, whereas another optimum gives a good description upstream. Locally optimal parametrizations therefore are a source of information to the model builder.

Identifiability
Identifiability concerns the question as to whether there exist several parameter values which correspond to the same model prediction. Are the model and the data sufficient to determine the parameter values uniquely? This question translates directly to the requirement to have a unique solution of the parameter estimation problem; Walter (1982). As will be illustrated, the global optimal set of parameters can consist of a line, a plane, or in general a manifold. In that case the parameters are called nonidentifiable. When a linear regression function

$$z(x, \beta) = \beta_1 + \beta_2 x$$

is fitted to the data of Figure 2.18, ordinary least squares (but also minimization of (2.12)) results in a unique solution, optimal parameter values (β_1, β_2). There is one best line through the points. Consider now the following model which is nonlinear in the parameters:

$$z(x, \beta) = \beta_1 \beta_2 x.$$

The multiplication of parameters sometimes appears when two linear regression relations are combined. This model corresponds with a line through the origin. The best line $y = constant \times x$ is uniquely defined; the parametrization, however, is not. All parameter values on the hyperbola $\beta_1 \beta_2 = $ CONSTANT give the same goodness of fit. The set of solutions of the optimization problem is a hyperbola. The parameters are nonidentifiable, i.e., cannot be determined individually. For the example this is relatively easy to see. For large models, analysis is necessary to determine the identifiability of the parameters. For the optimization problem this phenomenon is important. The number of optimal solutions is infinite.

Reliability

Often a researcher is more interested in exact parameter values than in how well the model fits the data. Consider an investigation on several treatments of a food product to influence the growth of bacteria in the product. A researcher measures the number of bacteria over time of several samples and fits growth models by estimating growth parameters. The researcher is interested in the difference of the estimated growth parameters with a certain reliability. So-called confidence regions are used for that.

2.7.1 Learning of neural nets seen as parameter estimation

A neural net can be considered as a model to translate input into output. It is doubtless that neural nets have been successfully applied in pattern recognition tasks; see, e.g., Haykin (1998). In the literature on Artificial Intelligence, a massive terminology has been introduced around this subject. Here we focus on the involved parameter estimation problem.

It is the task of a neural net to translate input x into output y. Therefore a neural net can be considered a model in the sense of Figure 2.17. Parameters are formed by so-called weights and biases and their values are used to tune the net. This can be seen as a large regression function. The tuning of the parameters, called learning in the appropriate terminology, can be considered as parameter estimation.

The net is considered as a directed graph with arcs and nodes. Every node represents a function which is a part of the total model. The output y of a node is a function of the weighted input $z = \sum w_i x_i$ and the so-called bias w_0, as sketched in Figure 2.19. So a node in the network has weights (on the input arcs) and a bias as corresponding parameters. The input z is transformed into output y by a so-called transformation function, usually the sigmoid or logistic transformation function:

$$y = \frac{1}{1 + \exp(w_0 - z)}.$$

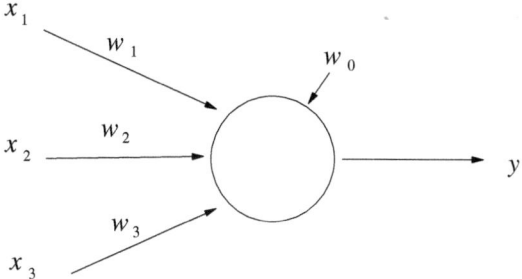

Fig. 2.19. One node of a neural net

This means that every individual node corresponds to a logistic regression function. The difference with application of such functions in growth and logit models is that the nodes are connected by arcs in a network. Therefore, the total net represents a large regression function. The parameters w_i can be estimated to describe relations as revealed by data as good as possible. For the illustration, a very small net is used as given in Figure 2.20. It consists of two so-called hidden nodes H_1 and H_2 and one output node y. Each node represents a logistic transformation function with three parameters, two on the incoming arcs and one bias. Parameters $w_1, w_2, \ldots, w_6$ correspond to weights on arcs and w_7, w_8 and w_9 are biases of the hidden and output node. The corresponding *regression function* is given by

$$y = \frac{1}{1 + \exp\left(w_9 - \dfrac{w_5}{1 + e^{w_7 - w_1 x_1 - w_2 x_2}} - \dfrac{w_6}{1 + e^{w_8 - w_3 x_1 - w_4 x_2}}\right)}.$$

The net is confronted with data. Usually the index p of "pattern" is used in the appropriate terminology. Input data x_p and output t_p (target) are said to be "fed to the network to train it." From a regression point of view, one wants parameters w_i to take values such that the predicted $y(x_p, w)$ fits the output observations t_p well according to a goodness of fit criterion such as

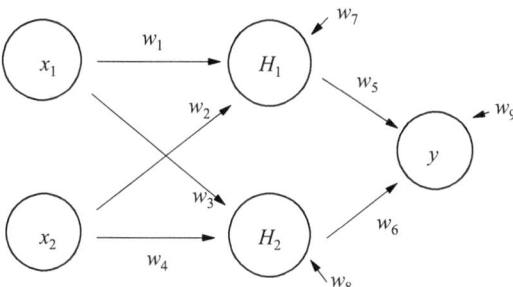

Fig. 2.20. Small neural net with two inputs, one output and two hidden nodes

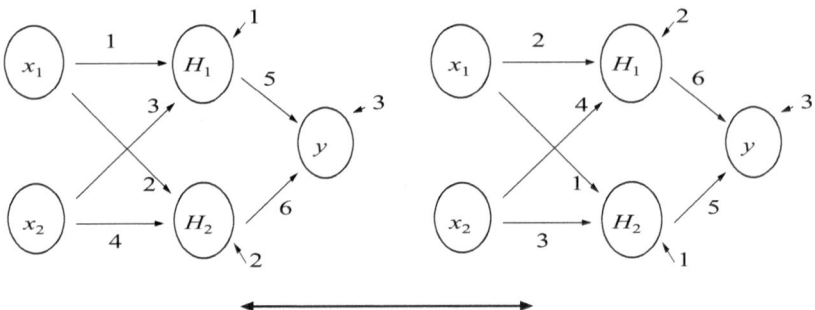

Fig. 2.21. Exchanging two hidden nodes in a neural net

$$f(w) = \sum_p (y_p - t_p)^2,$$

in which y_p is the regression function calculated for x_p and the weights w. Now we come across the symmetry property. The regression problem of neural nets is multiextremal due to its structure. After finding an optimal vector of weights for criterion $f(w)$, exchanging hidden nodes leads to reordering the parameters (or indices) and results in the same regression function and consequently to the same goodness of fit. The two simple nets in Figure 2.21 correspond to the same regression function. Parameter vector $(w_1, \ldots, w_9) = (1, 3, 2, 4, 5, 6, 1, 2, 3)$ gives the same regression function as parameter vector $(2, 4, 1, 3, 6, 5, 2, 1, 3)$. In general, a similar net with one output node and N hidden nodes has $N!$ optimal parameter vectors all describing the same input–output relation.

We have shown for this problem that the number of global optimal parametrizations is not necessarily infinite, but it grows more than exponentially with the number of hidden nodes due to an inherent symmetry in the optimization problem.

2.8 Summary and discussion points

This chapter taught us several aspects about modeling and optimization problems.

- In modeling optimization problems, one should distinguish clearly decision variables, given parameters and data, criteria and model structure.
- From the examples one can distinguish two types of problems from an optimization perspective.
    1. Black-box models: Parameter values are given to a model that returns the objective function value. Examples are simulation models like the centrifugal screen design and dynamic stochastic simulation such as the pumping rule and inventory control problems.

2. White-box case: explicit analytical expressions of the problem to be solved are assumed to be available. This was illustrated by the quadratic design cases.

- The dimension of a decision problem can blow up easily taking spatial and temporal aspects into the model, as illustrated by the economic models.
- Alternative optimal solutions may appear due to model structure. The alternative solutions may describe a complete lower-dimensional set, but also a finite number of alternatives that represent the same solution for the situation that has been modeled. The illustration was taken from parameter estimation problems.

2.9 Exercises

1. **Minimum enclosing sphere**
   Given a set of 10 points $\{p_1, \ldots, p_{10}\} \in \mathbb{R}^n$. The generic question is to find the (Chebychev) center c and radius r such that the maximum distance over the 10 points to center c is at its minimum value. This means, find a sphere around the set of points with a radius as small as possible.
   (a) Formulate the problem in vector notation (min-max problem).
   (b) Generate with a program or spreadsheet 10 points at random in $\mathbb{R}^2$.
   (c) Make a program or spreadsheet calculating the max distance given c.
   (d) Determine the center c with the aid of a solver.
   (e) Intuitively, does this problem have only one (local) optimum?

2. **Packing circles in a square**
   How to locate K points in a given square, such that the minimum distance (over all pairs) between them is as big as possible? Alternatively, this problem can be formulated in words as finding the smallest square around a given set of K equal size balls. Usually this problem is considered in a two-dimensional space. Figure 2.22 gives an optimal configuration for $K = 7$ spheres. Source: http://www.inf.u-szeged.hu/~pszabo/Pack.html
   (a) Formulate the problem in vector notation (max-min problem).

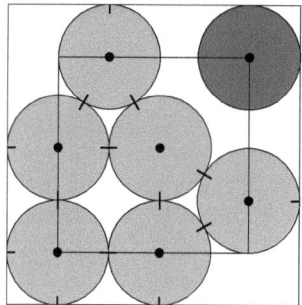

Fig. 2.22. Packing seven circles in a square

(b) Make a program or spreadsheet, such that for a configuration of K points determines the minimum distance between all point pairs.

(c) Make a program or spreadsheet that given a starting configuration of $K = 3$ points finds the maximum minimum distance between them in the unit box $[0, 1]^2$.

(d) Is there an intuitive argument to say that this problem has only one (local) optimum?

3. **Inventory control**

Consider an (s, Q)-policy as described in Section 2.3. As soon as the inventory is below level s, an order is placed of size Q which becomes available at the end of the next day. If there is not sufficient stock (inventory), the client is supplied the next day (back ordering) at additional cost. Take as cost data: inventory holding cost: 0.3 per unit per day, ordering cost 750 and back order cost of 3 per unit. The stochastic daily demand follows a triangular distribution with values between 0 and 800. This is the same as the addition $\xi = u_1 + u_2$ of two uniformly distributed random variables u_1 and u_2 between 0 and 400.

(a) Generate with a spreadsheet (or other program) 2000 daily demand data.

(b) Make a program or spreadsheet that determines the total costs given values for Q and s and the generated data.

(c) Determine good values for Q and s.

(d) Is the objective value (total costs) very sensitive to the values of s?

4. **Quadratic optimization**

Given property $y(x)$ as a result of a quadratic regression: $y(x) = -1 - 2x_1 - x_2 + x_1x_2 + x_1^2$. We would like to minimize $y(x)$ on a design space defined by $0 \le x_1 \le 3$ and $1 \le x_2 \le 4$. Determine whether $y(x)$ has several local optima on the design space.

5. **Regression**

Given four observations (x_i, y_i): $(0, 0), (\frac{1}{2}, 1), (1, 0)$ and $(\frac{3}{2}, -1)$. A corresponding regression model is given by $z(x, \alpha) = \sin(\alpha x)$. Determine the least squares regression function $f(\alpha)$ and evaluate $f(0)$ and $f(2\pi)$. What is the optimal value of α? Extend the exercise to the regression model $z(x, \alpha, \beta) = \sin(\alpha x) + \beta$.

6. **Marking via a neural net**

After 15 years of experience with the course "identifiability of strange species" a professor managed to compose an exam quickly. For giving marks, he found that in fact the result of the exam in the past only depended on two answers A and B of the candidates. A colleague developed a neural net for him which was fed the data of the exams of the past years to train it. Within one hour after the exam the professor had put into the net the answers A and B of the 50 students who participated. He calculated the marks and transferred them to the administration just

Table 2.1. Indicator values and corresponding marks

Indicator A	Indicator B	Mark	Indicator A	Indicator B	Mark
70.0	30.0	7.3	35.0	13.7	1.2
98.2	14.9	9.8	21.6	14.7	2.7
27.0	29.9	7.3	87.8	21.0	9.8
18.4	25.6	2.7	38.9	17.9	2.7
29.0	27.3	7.2	73.0	23.1	5.4
28.1	17.2	2.7	10.9	21.4	2.7
36.3	12.1	1.2	77.9	16.8	9.8
33.2	30.0	7.3	59.7	15.7	6.5
9.5	18.7	2.7	67.6	29.7	7.3
78.9	13.2	9.8	57.1	24.8	7.3
63.6	28.9	7.3	91.2	16.5	9.8
98.6	14.2	9.8	98.1	27.6	9.8
14.6	14.2	2.7	57.7	26.2	7.3
97.2	23.9	9.8	77.1	27.1	5.0
71.1	18.3	9.8	40.3	16.1	1.3
49.1	23.0	7.3	25.6	26.7	3.8
71.7	18.5	9.8	8.3	23.3	2.7
56.7	22.4	6.6	70.4	23.7	5.0
38.4	25.1	7.3	5.2	24.9	2.7
51.3	18.4	5.0	84.7	21.7	9.8
26.5	10.4	1.2	3.9	19.7	2.7
12.1	12.7	2.7	65.3	19.3	5.3
13.8	15.3	2.7	67.8	21.1	5.1
60.6	25.8	7.3	46.3	10.2	1.3
47.6	29.9	7.3	44.1	16.3	1.3

in time to catch the airplane for research on more strange species. The result is given in Table 2.1. The marks range from 0 to 10. The question is now to construct a neural net that matches the input and output of the exam results.

(a) Develop the simplest net without any nodes in the hidden layer, so that it consists of one output node. How many parameters does it have? What are the best values for the parameters in the sense that your results match the marks as good as possible (minimum least squares)?

(b) Add 1, 2 and 3 nodes to a hidden layer. What is the number of parameters in the network? Can some of the parameters be fixed without changing the function of the network?

(c) What is the closest you can get to the resulting marks (minimum least squares)?

3

NLP optimality conditions

3.1 Intuition with some examples

After an optimization problem has been formulated (or during the formulation), methods can be used to determine an optimal plan x^*. In the application of NLP algorithms, x^* is approximated iteratively. The user normally indicates how close an optimum should be approximated. We will discuss this in Chapter 4.

There are several ways to use software for nonlinear optimization. One can use development platforms like MATLAB. Modeling languages can be applied, such as GAMS and GINO. We found frequent use of them in economic studies, as sketched in Chapter 2. Also in spreadsheet environments, a so-called solver add-in on Excel, is used frequently. To get a feeling for the theories and examples here, one could use one of these programs. In the appendices an example can be found of the output of these programs.

The result of a method gives in practice an approximation of an optimal solution that fulfills the optimality conditions and that moreover gives information on sensitivity with respect to the data. First an example is introduced before going into abstraction and exactness.

Example 3.1. A classical problem in economics is the so-called utility maximization. In the two-goods case, x_1 and x_2 represent the amount of goods of type 1 and 2 and a utility function $U(x)$ is maximized. Given a budget (here 6 units) and prices for goods 1 and 2, with a value of 1 and 2, respectively, optimization problem (3.1) appears:

$$\max\{U(x) = x_1 x_2\}$$
$$x_1 + 2x_2 \leq 6 \qquad\qquad (3.1)$$
$$x_1, x_2 \geq 0.$$

To describe (3.1) in the terms of general NLP problem (1.1), one can define $f(x) = -U(x)$. Feasible area X is described by three inequalities $g_i(x) \leq 0$; $g_1(x) = x_1 + 2x_2 - 6$, $g_2(x) = -x_1$ and $g_3(x) = -x_2$.

E.M.T. Hendrix and B.G.-Tóth, *Introduction to Nonlinear and Global Optimization*, 31
Springer Optimization and Its Applications 37, DOI 10.1007/978-0-387-88670-1_3,
© Springer Science+Business Media, LLC 2010

In order to find the best plan, we should first define what an optimum, i.e., maximum or minimum, is. In Figure 1.1, the concept of a global and local optimum has been sketched. In words: a plan is called locally optimal when it is the best plan in its close environment. A plan is called globally optimal when it is the best plan in the total feasible area. In order to formalize this, it is necessary to define the concept of "close environment" in a mathematical way. The mathematical environment of x^* is given as a sphere (ball) with radius ϵ around x^*.

Definition 1. Let $x^* \in \mathbb{R}^n$, $\epsilon > 0$. Set $\{x \in \mathbb{R}^n \mid \|x - x^*\| < \epsilon\}$ is called an ϵ-environment of x^*, where $\|\|$ is a distance norm.

Definition 2. Function f has a minimum (or local minimum) over set X at x^* if there exists an ϵ-environment W of x^*, such that: $f(x) \geq f(x^*)$ for all $x \in W \cap X$.

In this case, vector x^* is called a minimum point (or local minimum point) of f. Function f has a global minimum in x^* if $f(x) \geq f(x^*)$ for all $x \in X$. In this case, vector x^* is called a *global minimum point*. The terminology strict minimum point is used when above $f(x) \geq f(x^*)$ is replaced by $f(x) > f(x^*)$ for $x \neq x^*$. In fact it means that x^* is a unique global minimum point. Note: For a maximization problem, in Definition 2 "minimum" is replaced by "maximum," the "$\geq$" sign by the "$\leq$" sign, and "$>$" by "$<$."

To determine the optimal plan for Example 3.1, the *contour* is introduced.

Definition 3. A contour of $f : \mathbb{R}^n \to \mathbb{R}$ of altitude h is defined as the set $\{x \in \mathbb{R}^n \mid f(x) = h\}$.

A contour can be drawn in a figure like lines of altitude on a map as long as $x \in \mathbb{R}^2$. Specifically in Linear Programming (LP), a contour $c_1 x_1 + c_2 x_2 = h$ is a line perpendicular to vector c. The contours of problem (3.1) are given in Figure 3.1.

Example 3.2. Similar to graphically solving LP, we can try to find the highest contour of $U(x)$, that has a point in common with feasible area X. The maximum point is $x^* = (3, 3/2)$, as indicated in Figure 3.1. The contours are so-called hyperbolas defined by $x_2 = h/x_1$.

It is not a coincidence that the optimum can be found where the contour of $U(x)$ touches the *binding* budget constraint $x_1 + 2x_2 \leq 6$, or abstractly $g_1(x) = 0$. As with many LP problems, the feasible area here consists of a so-called polytope. However, the optimum point x^* is not found in a vertex. For NLP problems, the optimum can be found in a vertex, on a line, on a plane or even in the interior of the feasible area, where no constraint is binding. So the number of binding constraints in the optimum point is not known in advance. We will come back to this phenomenon, but proceed now with the next example.

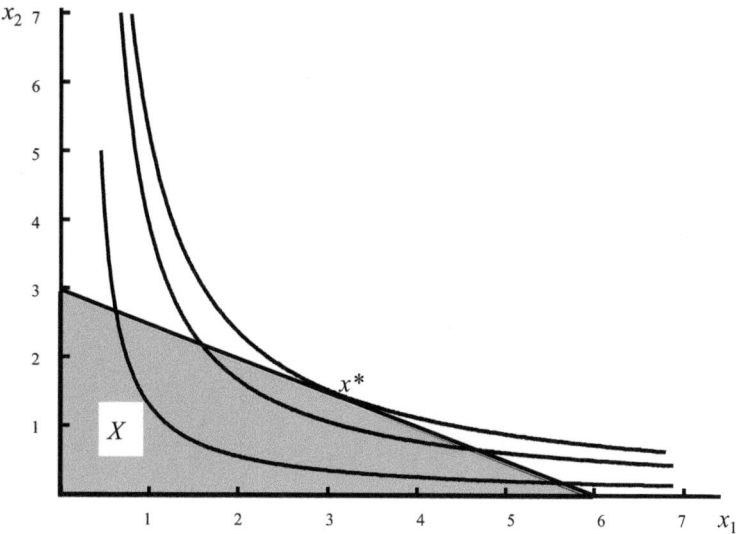

Fig. 3.1. Maximizing utility function $U(x) = x_1 x_2$

Example 3.3. In (3.1), the utility function is changed to $U(x) = x_1^2 + x_2^2$. The contours are now circles around the origin. In $x_1^* = (6,0)^T$ the global maximum point can be found; $f(x_1^*) = 36 \geq f(x) \; \forall x \in X$, corresponds to the highest feasible contour. The point $x_2^* = (0,3)^T$ is a local, nonglobal maximum point. That is, there exists an ϵ-environment of x_2^* (for instance for $\epsilon = 0.01$), such that all points situated in the intersection of this environment and X, have a lower objective function value. The point $x_3^* = \frac{1}{5}(6,12)^T$ is a point where a contour touches a binding constraint. In contrast to Example 3.2, such a point is not a maximum point.

What can be derived from Example 3.3? An optimum point possibly can be found in a vertex of the feasible area. In contrast to Linear Programming, there can exist points that are local nonglobal optima. A last example in this section is discussed.

Example 3.4. The concept of investing money in a number of goods can also be found in the decisions of investing in funds, so-called portfolio selection. Traditionally a trade-off has to be made between return on investment and risk. In the classical E, V model introduced by Markowitz (1959), a number of funds, so-called portfolios, are given. Every portfolio has an expected return μ_j and variance σ_j^2, $j = 1, \ldots, n$. A fixed amount, say 100€, should be invested in funds, such that the total expected return is as high as possible and the risk (variance) as low as possible. Let x_j be the amount invested in portfolio j; then

$$\sum x_j = 100, \; x_j \geq 0 \quad j = 1, \ldots, n$$

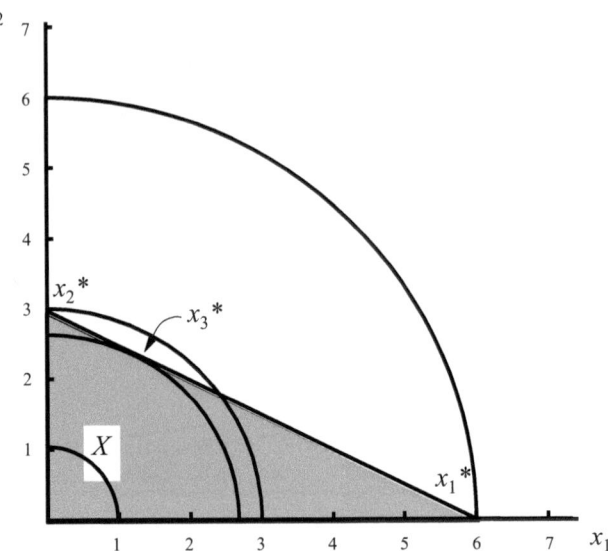

Fig. 3.2. Several optima maximizing utility $U(x) = x_1^2 + x_2^2$

Expected return $E = \sum x_j \mu_j$

Variance of return $V = \sum x_j^2 \sigma_j^2 + 2 \sum_{i=1}^n \sum_{j=i+1}^n \sigma_{ij} x_i x_j$.

The term σ_{ij} is the so-called covariance between funds i and j. This can be positive (sunglasses and sunscreen), negative (sunglasses and umbrellas) and zero; the funds have no relation. The investor wants to maximize E and at the same time minimize V. This is a so-called multiobjective problem. In this case one can generate a trade-off curve between maximizing E and minimizing V. The simplest case, similar to the examples discussed, is when there are two funds. Substituting the binding budget constraint $x_2 + x_1 = 100 \rightarrow x_2 = 100 - x_1$, a problem in one decision variable appears that can be analyzed relatively easily. Given two funds with $\mu_1 = 1$, $\sigma_1 = 1$, $\mu_2 = 2$, $\sigma_2 = 2$ and $\sigma_{12} = 0$. The variance $V = x_1^2 + 4x_2^2$ by substituting $x_2 = 100 - x_1$ becomes $V(x_1) = 5x_1^2 - 800x_1 + 40\,000$.

This describes a parabola with a minimum at $x_1^* = 80$. An investor who wants to minimize the risk has an optimal plan $x^* = (80, 20)$ where $E = 120$ and $V = 8\,000$. This optimal plan is not a vertex of the feasible area. Maximizing the expected return E gives the optimal solution $x^* = (0, 100)$ where $E = 200$ and $V = 40\,000$. This is a vertex of the feasible area. In practice an investor will make a trade-off between expected return and acceptable risk. In this specific case of two funds, if one makes a graph with E and V on the axes as a function of x_1, the trade-off line of Figure 3.3 appears. In the general Markowitz model, points on this curve are derived by maximizing an

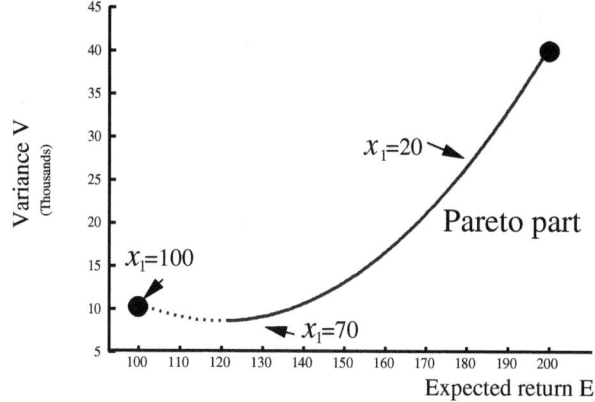

Fig. 3.3. Trade-off between variance and expected return

objective function $f(x) = E - \beta V$, where β is called a risk aversion parameter. For every value of β, a so-called efficient point or Pareto point appears.

Determination of the optimal plan for varying parameter values is called *parametric programming*. In LP this leads to piecewise linear curves. As can be observed from Example 3.4, in NLP this can be a smooth nonlinear curve.

Exercising with simple examples based on graphical analysis gives several insights:

- An optimum point cannot always be found in a vertex of the feasible area.
- The point where a contour touches a constraint is a special point.
- Local, nonglobal optimum points may exist.
- Changes in parameter values (parametric programming) may lead to non-linear curves.

The rest of the chapter concerns the formalization of these insights and the question as to in what situations what phenomena appear.

3.2 Derivative information

Some terminology is introduced for the formalization. An objective function $f : \mathbb{R}^n \to \mathbb{R}$ is analyzed. In the first section, the contour $\{x \in \mathbb{R}^n \mid f(x) = h\}$ has already been introduced. A *graph* $\{(x, y) \in \mathbb{R}^{n+1} \mid f(x) = y\}$ and *level set* $\{x \in \mathbb{R}^n \mid f(x) \leq h\}$ are further concepts. The level set is also sometimes called the sublevel set. The graphical perception is limited to $\mathbb{R}^3$. A graph as such can be seen from the point of view of a landscape. In the development of algorithms as well as theory, many concepts can be derived by cross-cutting a function $f : \mathbb{R}^n \to \mathbb{R}$ considering a function φ_r of one variable starting in a fixed point x and looking in the direction r:

$$\varphi_r(\lambda) = f(x + \lambda r). \tag{3.2}$$

3.2.1 Derivatives

The notion of a derivative is first considered for a function in one variable and then via (3.2) extended to functions of several variables. The derivative of $f : \mathbb{R} \to \mathbb{R}$ in the point x is defined as

$$f'(x) = \lim_{h \to 0} \frac{f(x+h) - f(x)}{h} \tag{3.3}$$

whenever this limit exists. For instance, for $f(x) = \sqrt{|x|}$ this limit does not exist for $x = 0$. The use of absolute value and the max-min structure in an optimization model as illustrated in Chapter 2, in practice causes an objective function not to be differentiable everywhere. Another important concept in mathematical literature is that of *continuously differentiable*, i.e., the derivative function is a continuous function. This is mainly a theoretical concept. It requires some imagination to come up with a function that is differentiable, but not continuously differentiable. In textbooks one often finds the example: $f(x) = x^2 \sin \frac{1}{x}$ when $x \neq 0$ and $f(x) = 0$ when $x = 0$. Exercising with limits shows that this function is differentiable for $x = 0$, but that the derivative is not continuous.

Algorithms often make use of the value of the derivative. Unless computer programs can manipulate formulas (so-called automatic differentiation), the user has to feed the program explicitly with the formulas of the derivatives. Usually, however, a computer package makes use of so-called *numerical differentiation*. In many cases outlined in Chapter 2, the calculation of a function value is the result of a long calculation process and an expression for the derivative is not available. A numerical approximation of the derivative is determined for instance by the *progressive* or *forward difference* approximation

$$f'(x) \approx \frac{f(x+h) - f(x)}{h} \tag{3.4}$$

by taking a small step (e.g., $h = 10^{-5}$) forward. Numerical errors can occur that, without going into detail, are smaller in the *central difference approximation*

$$f'(x) \approx \frac{f(x+h) - f(x-h)}{2h}. \tag{3.5}$$

3.2.2 Directional derivative

By considering functions of several variables from a one-dimensional perspective via (3.2), the notion of *directional derivative* $\varphi'_r(0)$ appears, depending on point x and direction r:

$$\varphi'_r(0) = \lim_{h \to 0} \frac{f(x+hr) - f(x)}{h}. \tag{3.6}$$

This notion is relevant for the design of search algorithms as well as for the test on optimality. When an algorithm has generated a point x, a direction r can be classified according to

r with $\varphi_r'(0) < 0$: descent direction,
r with $\varphi_r'(0) > 0$: ascent direction,
r with $\varphi_r'(0) = 0$: direction in which f does not increase nor decrease, it is situated in the tangent plane of the contour.

For algorithms looking for the minimum of a differentiable function, in every generated point, the descent directions are of interest. For the test of whether a certain point x is a minimum point of a differentiable function the following reasoning holds. In a minimum point x^* there exists no search direction r that points into the feasible area and is also a descent direction, $\varphi_r'(0) < 0$. Derivative information is of importance for testing optimality. The test of a set of possible search directions requires the notion of *gradient* which is introduced now.

3.2.3 Gradient

Consider unit vector e_j with a 1 for element j and 0 for the other elements. Using in (3.6) e_j for the direction r gives the so-called partial derivative:

$$\frac{\partial f}{\partial x_j}(x) = \lim_{h \to 0} \frac{f(x + he_j) - f(x)}{h}. \tag{3.7}$$

The vector of partial derivatives is called the gradient

$$\nabla f(x) = \left(\frac{\partial f}{\partial x_1}(x), \frac{\partial f}{\partial x_2}(x), \dots, \frac{\partial f}{\partial x_n}(x) \right)^T. \tag{3.8}$$

Example 3.5. Let $f : \mathbb{R}^n \to \mathbb{R}$ be a linear function $f(x) = c^T x$. The partial derivatives of f in x with respect to x_j are

$$\frac{\partial f}{\partial x_j}(x) = \lim_{h \to 0} \frac{f(x + he_j) - f(x)}{h}$$

$$= \lim_{h \to 0} \frac{c^T(x + he_j) - c^T x}{h} = \lim_{h \to 0} \frac{hc^T e_j}{h} = c^T e_j = c_j.$$

Gradient $\nabla f(x) = (c_1, \dots, c_n)^T = c$ for linear functions does not depend on x.

Example 3.6. Consider again utility optimization problem (3.1). Utility function $U(x) = x_1 x_2$ has gradient $\nabla U(x) = (x_2, x_1)^T$. The gradient ∇U is depicted for several plans x in Figure 3.4(a). The arrow $\nabla U(x)$ is perpendicular to the contour; this is not a coincidence. In Figure 3.4(b), contours of another function can be found from the theory of utility maximization. It

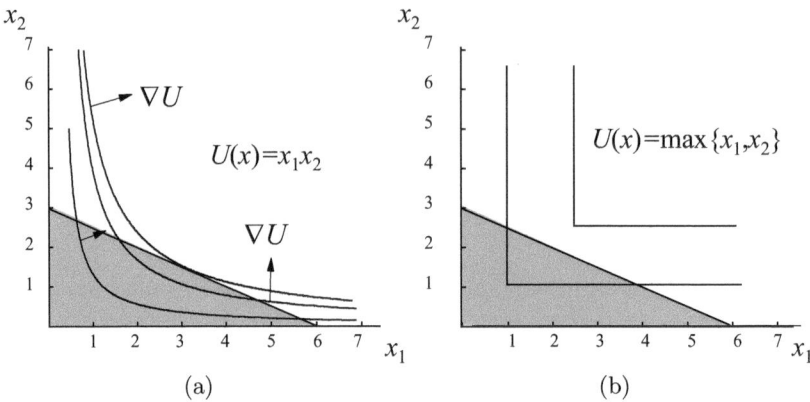

Fig. 3.4. Two utility functions and their contours

concerns the notion of complementary goods. A classical example of complementary goods is where x_1 is the number of right shoes, x_2 is the number of left shoes and $U(x) = \min\{x_1, x_2\}$ the number of pairs of shoes, that seems to be maximized by some individuals. This utility function is not differentiable everywhere. Consider the graph over the line $x + \lambda r$, with $x = (0, 1)^T$ and $r = (2, 1)^T$. The function is then

$$\varphi_r(\lambda) = U\left(\begin{pmatrix} 0 \\ 1 \end{pmatrix} + \lambda \begin{pmatrix} 2 \\ 1 \end{pmatrix} \right).$$

Note that for any direction r, $\varphi_r(\lambda)$ is a parabola when $U(x) = x_1 x_2$ and a piecewise linear curve when $U(x) = \min\{x_1, x_2\}$.

If f is continuously differentiable in x, the directional derivative is

$$\varphi_r'(0) = r^T \nabla f(x). \tag{3.9}$$

This follows from the chain rule for differentiating a composite function with respect to λ:

$$\varphi_r'(\lambda) = \frac{d}{d\lambda} f(x + \lambda r) = r_1 \frac{\partial}{\partial x_1} f(x + \lambda r) + \cdots + r_n \frac{\partial}{\partial x_n} f(x + \lambda r)$$
$$= r^T \nabla f(x + \lambda r).$$

Using (3.9), the classification of search directions toward descent and ascent directions becomes relatively easy. For a descent direction r, $r^T \nabla f(x) = \varphi_r'(0) < 0$ holds, such that r makes a sharp angle with $\nabla f(x)$. Directions for which $r^T \nabla f(x) > 0$, are directions where f increases.

3.2.4 Second-order derivative

The second-order derivative in the direction r is defined similarly. However, the notation gets more complicated, as every partial derivative $\frac{\partial f}{\partial x_j}$ has derivatives with respect to $x_1, x_2, \ldots, x_n$. All derivatives can be summarized in the so-called Hesse matrix $H(x)$ with elements $h_{ij} = \frac{\partial^2 f}{\partial x_i x_j}(x)$. The matrix is named after the German mathematician Ludwig Otto Hesse (1811–1874). In honor of this man, in this text the matrix is not called Hessian (as usual), but Hessean:

$$H_f(x) = \begin{pmatrix} \frac{\partial^2 f}{\partial x_1 \partial x_1}(x) & \cdots & \frac{\partial^2 f}{\partial x_1 \partial x_n}(x) \\ \vdots & \ddots & \vdots \\ \frac{\partial^2 f}{\partial x_n \partial x_1}(x) & \cdots & \frac{\partial^2 f}{\partial x_n \partial x_n}(x) \end{pmatrix}.$$

Example 3.7. Let $f : \mathbb{R}^n \to \mathbb{R}$ be defined by $f(x) = x_1^3 x_2 + 2x_1 x_2 + x_1$:

$$\frac{\partial f}{\partial x_1}(x) = 3x_1^2 x_2 + 2x_2 + 1 \qquad \frac{\partial f}{\partial x_2}(x) = x_1^3 + 2x_1$$

$$\frac{\partial^2 f}{\partial x_1 \partial x_1}(x) = 6x_1 x_2 \qquad \frac{\partial^2 f}{\partial x_1 \partial x_2}(x) = 3x_1^2 + 2$$

$$\frac{\partial^2 f}{\partial x_2 \partial x_1}(x) = 3x_1^2 + 2 \qquad \frac{\partial^2 f}{\partial x_2 \partial x_2}(x) = 0$$

$$\nabla f(x) = \begin{pmatrix} 3x_1^2 x_2 + 2x_2 + 1 \\ x_1^3 + 2x_1 \end{pmatrix} \qquad H_f(x) = \begin{pmatrix} 6x_1 x_2 & 3x_1^2 + 2 \\ 3x_1^2 + 2 & 0 \end{pmatrix}.$$

In the rest of this text, the suffix f of the Hessean will only be used when it is not clear from the context which function is meant. The Hessean in this example is a symmetric matrix. It can be shown that the Hessean is symmetric if f is *twice continuously differentiable*. For the optimality conditions we are interested in the second-order derivative $\varphi_r''(\lambda)$ in the direction r. Proceeding with the chain rule on (3.9) results in

$$\varphi_r''(\lambda) = r^T H(x + \lambda r) r \tag{3.10}$$

if f is twice continuously differentiable.

Example 3.8. The function $U(x) = x_1 x_2$ implies

$$\nabla U(x) = \begin{pmatrix} x_2 \\ x_1 \end{pmatrix} \quad \text{and} \quad H(x) = \begin{pmatrix} 0 & 1 \\ 1 & 0 \end{pmatrix}.$$

The Hessean is independent of the point x. Consider the one-dimensional function $\varphi_r(\lambda) = U(x + \lambda r)$ with $x = (0, 0)^T$. In the direction $r = (1, 1)^T$ is a parabola $\varphi_r(\lambda) = \lambda^2$,

$$\varphi_r'(\lambda) = r^T \nabla U(x + \lambda r) = (1, 1) \begin{pmatrix} \lambda \\ \lambda \end{pmatrix} = 2\lambda \text{ and}$$

$$\varphi_r''(\lambda) = r^T H(x + \lambda r) r = (1, 1) \begin{pmatrix} 0 & 1 \\ 1 & 0 \end{pmatrix} \begin{pmatrix} 1 \\ 1 \end{pmatrix} = 2.$$

In the direction $r = (1, -1)^T$ is $\varphi_r(\lambda) = -\lambda^2$ (parabola with maximum) such that

$$\varphi'_r(\lambda) = (1, -1)\begin{pmatrix} -\lambda \\ \lambda \end{pmatrix} = -2\lambda \text{ and}$$

$$\varphi''_r(\lambda) = (1, -1)\begin{pmatrix} 0 & 1 \\ 1 & 0 \end{pmatrix}\begin{pmatrix} 1 \\ -1 \end{pmatrix} = -2.$$

In $x = (0, 0)^T$ there are directions in which x is a minimum point and there are directions in which x is a maximum point. Such a point x is called a *saddle point*.

Definition 4. Point x is a saddle point if there exist directions r and s for which $\varphi_r(\lambda) = f(x + \lambda r)$ has a minimum in $\lambda = 0$ and $\varphi_s(\lambda) = f(x + \lambda s)$ has a maximum in $\lambda = 0$.

3.2.5 Taylor

The first- and second-order derivatives play a role in the so-called mean value theorem and Taylor's theorem. Higher-order derivatives that are usually postulated in the theorem of Taylor are left out here.

The mean value theorem says that for a differentiable function between two points a and b a point ξ exists where the derivative has the same value as the slope between $(a, f(a))$ and $(b, f(b))$.

Theorem 3.1. Mean value theorem. *Let $f : \mathbb{R} \to \mathbb{R}$ be continuous on the interval $[a, b]$ and differentiable on (a, b), then $\exists \, \xi$, $a \leq \xi \leq b$, such that*

$$f'(\xi) = \frac{f(b) - f(a)}{b - a}. \tag{3.11}$$

As a consequence, considered from a point x_1, the function value $f(x)$ is

$$f(x) = f(x_1) + f'(\xi)(x - x_1). \tag{3.12}$$

So $f(x)$ equals $f(x_1)$ plus a residual term that depends on the derivative in a point in between x and x_1 and the distance between x and x_1. The residual or error idea can also be found in Taylor's theorem. For a twice-differentiable function, (3.12) can be extended to

$$f(x) = f(x_1) + f'(x_1)(x - x_1) + \frac{1}{2}f''(\xi)(x - x_1)^2. \tag{3.13}$$

It tells us that $f(x)$ can be approximated by the tangent line through x_1 and that the error term is determined by the second-order derivative in a point ξ in between x and x_1. The tangent line $f(x_1) + f'(x_1)(x - x_1)$ is called the first-order Taylor approximation. The equivalent terminology for functions of several variables can be derived from the one-dimensional cross-cut function φ_r given in (3.2).

We consider vector x_1 as a fixed point and do a step into direction r, such that $x = x_1 + r$; given $\varphi_r(\lambda) = f(x_1 + \lambda r)$, consider $\varphi_r(1) = f(x)$. The mean value theorem gives

$$f(x) = \varphi_r(1) = \varphi_r(0) + \varphi_r'(\xi) = f(x_1) + r^T \nabla f(\theta) = f(x_1) + (x - x_1)^T \nabla f(\theta),$$

where θ is a vector in between x_1 and x. The *first-order Taylor approximation* becomes

$$f(x) \approx f(x_1) + (x - x_1)^T \nabla f(x_1). \tag{3.14}$$

This line of reasoning via (3.10) results in Taylor's theorem (second order)

$$\begin{aligned} f(x) = \varphi_r(1) &= \varphi_r(0) + \varphi_r'(0) + \tfrac{1}{2}\varphi_r''(\xi) \\ &= f(x_1) + (x - x_1)^T \nabla f(x_1) + \tfrac{1}{2}(x - x_1)^T H(\theta)(x - x_1), \end{aligned} \tag{3.15}$$

where θ is a vector in between x_1 and x. The second-order Taylor approximation appears when in (3.15) θ is replaced by x_1. The function of equation (3.15) is a so-called quadratic function. In the following section we will first focus on this type of functions.

Example 3.9. Let $f(x) = x_1^3 x_2 + 2x_1 x_2 + x_1$, see Example 3.7. The first-order Taylor approximation of $f(x)$ around 0 is

$$f(x) \approx f(0) + x^T \nabla f(0)$$

$$\nabla f(0) = \begin{pmatrix} 1 \\ 0 \end{pmatrix}$$

$$f(x) \approx 0 + (x_1, x_2) \begin{pmatrix} 1 \\ 0 \end{pmatrix} = x_1.$$

In Section 3.4, the consequence of first- and second-order derivatives with respect to optimality conditions is considered. First the focus will be on the specific shape of quadratic functions at which we arrived in equation (3.15).

3.3 Quadratic functions

In this section, we focus on a special class of functions and their optimality conditions. In the following sections we expand on this toward general smooth functions. At least for any smooth function the second-order Taylor equation (3.15) is valid, which is a quadratic function. In general a quadratic function $f : \mathbb{R}^n \to \mathbb{R}$ can be written as

$$f(x) = x^T A x + b^T x + c, \tag{3.16}$$

where A is a symmetric $n \times n$ matrix and b an n vector. Besides constant c and linear term $b^T x$ (3.16) has a so-called quadratic form $x^T A x$. Let us first consider this quadratic form, as has already been exemplified in Example 3.8:

$$x^T A x = \sum_{i=1}^{n} \sum_{j=1}^{n} a_{ij} x_i x_j \qquad (3.17)$$

or alternatively as was written in the portfolio example, Example 3.4:

$$x^T A x = \sum_{i=1}^{n} a_{ii} x_i^2 + 2 \sum_{i=1}^{n} \sum_{j=i+1}^{n} a_{ij} x_i x_j. \qquad (3.18)$$

Example 3.10. Let $A = \begin{pmatrix} 2 & 1 \\ 1 & 1 \end{pmatrix}$, then $x^T A x = 2x_1^2 + x_2^2 + 2x_1 x_2$.

The quadratic form $x^T A x$ determines whether the quadratic function has a maximum, minimum or neither of them. The quadratic form has a value of 0 in the origin. This would be a minimum of $x^T A x$, if $x^T A x \geq 0$ for all $x \in \mathbb{R}$ or similarly in the line of the cross-cut function $\varphi_r(\lambda) = f(0 + \lambda r)$, walking in any direction r would give a nonnegative value:

$$(0 + \lambda r)^T A (0 + \lambda r) = \lambda^2 r^T A r \geq 0 \quad \forall r. \qquad (3.19)$$

We will continue this line of thinking in the following section. For quadratic functions, it brings us to introduce a useful concept.

Definition 5. Let A be a symmetric $n \times n$ matrix. A is called positive definite if $x^T A x > 0$ for all $x \in \mathbb{R}^n$, $x \neq 0$. Matrix A is called positive semidefinite if $x^T A x \geq 0$ for all $x \in \mathbb{R}^n$. The notion of negative (semi)definite is defined analogously. Matrix A is called indefinite if vectors x_1 and x_2 exist such that $x_1^T A x_1 > 0$ and $x_2^T A x_2 < 0$.

The status of matrix A with respect to positive, negative definiteness or indefiniteness determines whether quadratic function $f(x)$ has a minimum, maximum or neither of them. The question is of course how to check the status of A. One can look at the eigenvalues of a matrix. It can be shown that for the quadratic form

$$\mu_1 \|x\|^2 \leq x^T A x \leq \mu_n \|x\|^2$$

where μ_1 is the smallest and μ_n the highest eigenvalue of A and $\|x\|^2 = x^T x$. This means that for a positive definite matrix A all eigenvalues are positive and for a negative definite matrix all eigenvalues are negative.

Theorem 3.2. *Let A be a symmetric $n \times n$ matrix. A is positive definite $\Longleftrightarrow$ all eigenvalues of A are positive.*

Moreover, the corresponding eigenvectors are orthogonal to the contours of $f(x)$. The eigenvalues of A can be determined by finding those values of μ for which $Ax = \mu x$ or equivalently $(A - \mu E)x = 0$ such that the determinant $|A - \mu E| = 0$. Let us look at some examples.

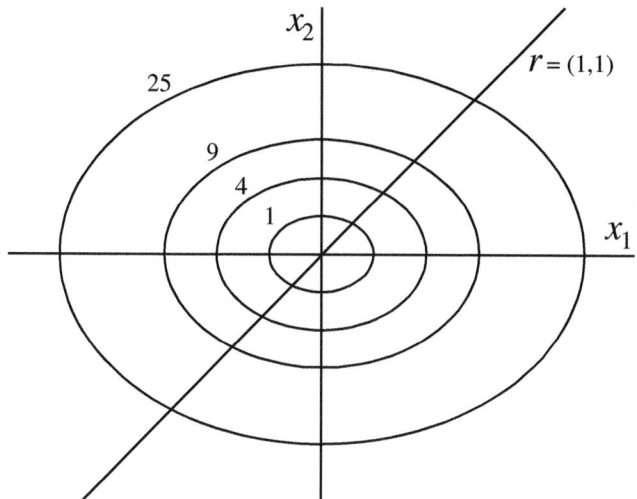

Fig. 3.5. Contours of $f(x) = x_1^2 + 2x_2^2$

Example 3.11. Consider $A = \begin{pmatrix} 1 & 0 \\ 0 & 2 \end{pmatrix}$ such that $f(x) = x_1^2 + 2x_2^2$. The corresponding contours as sketched in Figure 3.5 are ellipsoids. The eigenvalues can be found on the diagonal of A and are 1 and 2 with corresponding eigenvectors $r_1 = (1,0)^T$ and $r_2 = (0,1)^T$. Following cross-cut functions from the origin according to (3.19) gives positive parabolas $\varphi_{r_1}(\lambda) = \lambda^2$ and $\varphi_{r_2}(\lambda) = 2\lambda^2$. Walking into direction $r = (1,1)^T$ also results in a positive parabola, but as depicted, the corresponding line is not orthogonal to the contours of $f(x)$.

In Example 3.8 we have seen already a case of an indefinite quadratic form. In some directions the parabola curves downward and in some directions it curves upward. We consider here one example where this is less obvious.

Example 3.12. Consider $A = \begin{pmatrix} 3 & 4 \\ 4 & -3 \end{pmatrix}$ such that $f(x) = 3x_1^2 - 3x_2^2 + 8x_1x_2$
The corresponding contours are sketched in Figure 3.6. The eigenvalues of A can be determined by finding those values of μ for which

$$|A - \mu E| = \begin{vmatrix} 3 - \mu & 4 \\ 4 & -3 - \mu \end{vmatrix} = 0 \quad \rightarrow \quad \mu^2 - 25 = 0 \qquad (3.20)$$

such that the eigenvalues are $\mu_1 = 5$ and $\mu_2 = -5$; A is indefinite. The eigenvector can be found from $Ar = \mu r \rightarrow (A - \mu E)r = 0$. In this example they are any multiple of $r_1 = \frac{1}{\sqrt{5}}(2,1)^T$ and $r_2 = \frac{1}{\sqrt{5}}(1,-2)^T$. The corresponding lines, also called axes, are given in Figure 3.6. In the direction of r_1, $\varphi_{r_1}(\lambda) = 5\lambda^2$ a positive parabola. In the direction of r_2 we have a negative parabola. Specifically in the direction $r = (1,3)$, $f(x)$ is constant.

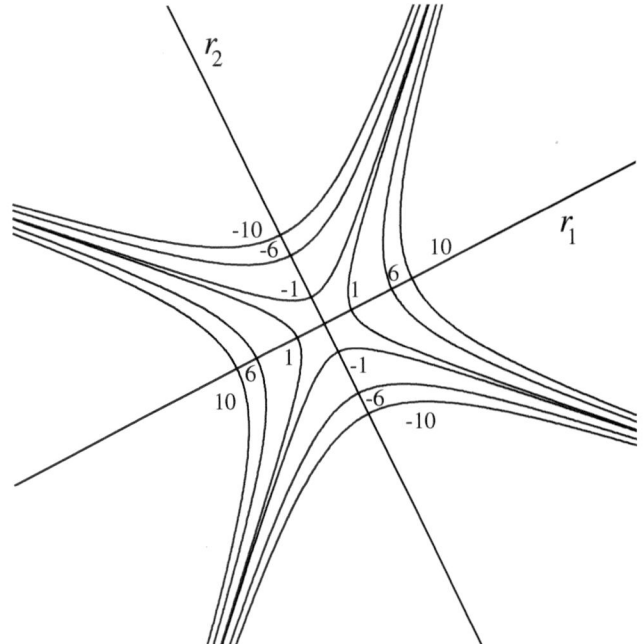

Fig. 3.6. Contours of $f(x) = 3x_1^2 - 3x_2^2 + 8x_1x_2$

When the linear term $b^T x$ is added to the quadratic form, the center of the contours is shifted toward

$$x^* = -\frac{1}{2}A^{-1}b \qquad (3.21)$$

where x^* can only be determined if the columns of A are linearly independent. In this case (3.16) can be written as

$$f(x) = x^T Ax + b^T x + c = (x - x^*)^T A(x - x^*) + constant, \qquad (3.22)$$

where $constant = c - \frac{1}{4}b^T A^{-1}b$. Combining Definition 5 with Equation (3.22) gives that apparently x^* is a minimum point if A is positive semidefinite and a maximum point if A is negative semidefinite.

The derivative information of quadratic functions is typically linear. The gradient of quadratic (3.22) is given by

$$\nabla f(x) = 2Ax + b. \qquad (3.23)$$

Note that point x^* is a point where the gradient is the zero vector, a so-called stationary point. The Hessean of a quadratic function is constant:

$$H(x) = 2A. \qquad (3.24)$$

Another typical observation can be made going back to the mean value theorem of Section 3.2. There exists a vector θ in between points a and d such that

$$f(d) = f(a) + \nabla f(\theta)^T (d - a). \tag{3.25}$$

In general, the exact location of θ is unknown. However, one can check that for quadratic functions θ is exactly in the middle; $\theta = \frac{1}{2}(a + d)$.

Example 3.13. Consider $A = \begin{pmatrix} 1 & 0 \\ 0 & 2 \end{pmatrix}$, $b = \begin{pmatrix} 2 \\ 4 \end{pmatrix}$, $f(x) = x_1^2 + 2x_2^2 + 2x_1 + 4x_2$. The center of Figure 3.5 is now determined by

$$x^* = -\frac{1}{2}A^{-1}b = -\frac{1}{2}\begin{pmatrix} 1 & 0 \\ 0 & \frac{1}{2} \end{pmatrix}\begin{pmatrix} 2 \\ 4 \end{pmatrix} = -\begin{pmatrix} 1 \\ 1 \end{pmatrix} \tag{3.26}$$

and *constant* $= -\frac{1}{4}b^T A^{-1}b = -\frac{1}{4}(2,4)^T \begin{pmatrix} 1 & 0 \\ 0 & \frac{1}{2} \end{pmatrix}\begin{pmatrix} 2 \\ 4 \end{pmatrix} = -3$ such that $f(x)$ can be written as $f(x) = x_1^2 + 2x_2^2 + 2x_1 + 4x_2 = (x_1 + 1)^2 + 2(x_2 + 1)^2 - 3$.

3.4 Optimality conditions, no binding constraints

An optimum point is determined by the behavior of the objective function in all feasible directions. If $f(x)$ is increasing from x^* in all feasible directions r, then x^* is a minimum point. The feasibility of directions is determined by the constraints that are binding in x^*. Traditionally, two situations are distinguished:

1. There are no binding constraints in x^*, x^* is an interior point of X. We will deal with that in this section.
2. There are binding constraints, x^* is situated at the boundary of X. We deal with that in Section 3.5.

The same line is followed as in Section 3.2 starting with one-dimensional functions via the cross-cut functions $\varphi_r(\lambda)$ to functions of several variables. Mathematical background and education often give the principle of putting derivatives to zero, popularly called "finding an analytical solution." The mathematical background of this principle is sketched here and commented.

3.4.1 First-order conditions

The general conditions are well described in the literature such as Bazaraa et al. (1993). We describe here some properties for f continuously differentiable. Considering a minimum point x^* of a one-dimensional function, gives via the definition of derivative

$$f'(x^*) = \lim_{x \to x^*} \frac{f(x) - f(x^*)}{x - x^*}. \tag{3.27}$$

The numerator of the quotient is nonnegative (x^* is a minimum point) and the denominator is either negative or positive depending on x approaching x^* from below or from above. So the limit in (3.27) can only exist if $f'(x^*) = 0$. More-dimensional functions follow the same property with the additional complication that the directional derivative

$$\varphi_r'(0) = \lim_{h \to 0} \frac{f(x^* + hr) - f(x^*)}{h} = r^T \nabla f(x^*) \tag{3.28}$$

depends on direction r. The directional derivative being zero for all possible directions, $r^T \nabla f(x^*) = 0 \ \forall r$ implies $\nabla f(x^*) = 0$.

A point x with $\nabla f(x) = 0$ is called a *stationary point*. Finding one (or all) stationary points results in a set of n equalities and n unknowns and in general cannot be easily solved. Moreover, a stationary point can be:

- A minimum point; $f(x) = x^2$ and $x = 0$
- A maximum point; $f(x) = -x^2$ and $x = 0$
- A point of inflection; $f(x) = x^3$ and $x = 0$
- A saddle point, i.e., in some directions a maximum point and in others a minimum point (Example 3.8)
- Combination of inflection, minimum or maximum point in different directions.

The variety is illustrated by Example 3.14.

Example 3.14. Let $f(x) = (x_1^3 - 1)^2 + (x_2^3 - 1)^2$. The contours of f are depicted in Figure 3.7 having decreasing function values around a minimum point in the positive orthant. The gradient of f is $\nabla f(x) = \begin{pmatrix} 6x_1^2(x_1^3 - 1) \\ 6x_2^2(x_2^3 - 1) \end{pmatrix}$. The stationary points can easily be found: $\nabla f(x) = 0$ gives $6x_1^2(x_1^3 - 1) = 0$ and $6x_2^2(x_2^3 - 1) = 0$. The stationary points are $(0, 0)$; $(1, 1)$; $(1, 0)$ and $(0, 1)$. The function value f in $(1, 1)$ equals zero and it is easy to see that $f(x) > 0$ for all other points. So point $(1, 1)$ is a global minimum point. The other stationary points $(0, 0)$; $(1, 0)$ and $(0, 1)$ are situated on a contour such that in their direct environment there exist points with a higher function value as well as points with a lower function value; they are neither minimum nor maximum points.

3.4.2 Second-order conditions

The assumption is required that f is twice continuously differentiable. Now Taylor's theorem can be used. Given a point x^* with $f'(x^*) = 0$, then (3.13) tells us that

$$f(x) = f(x^*) + \frac{1}{2} f''(\xi)(x - x^*)^2. \tag{3.29}$$

Whether x^* is a minimum point is determined by the sign of $f''(\xi)$ in the environment of x^*. If $f'(x^*) = 0$ and $f''(\xi) \geq 0$ for all ξ in an environment,

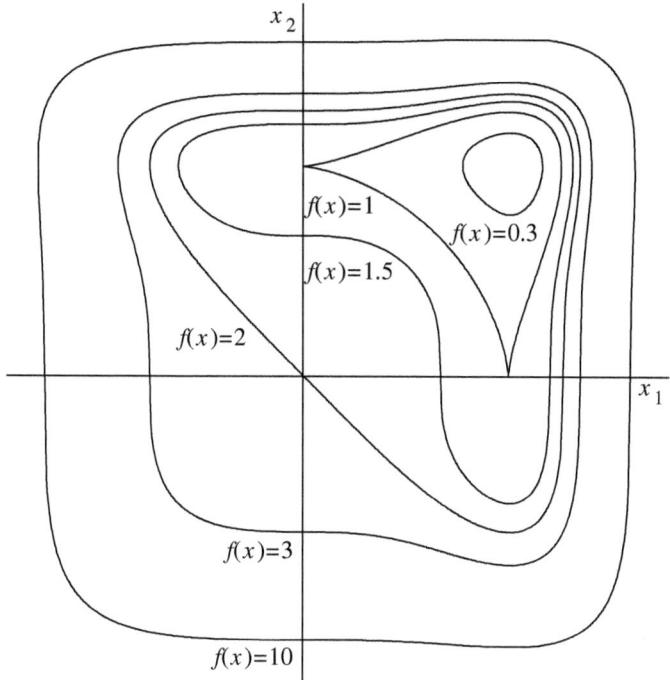

Fig. 3.7. Contours of $f(x) = (x_1^3 - 1)^2 + (x_2^3 - 1)^2$

then x^* is a minimum point. When f'' is a continuous function and $f''(x^*) > 0$, then there exists an environment of x^* such that for all points in that environment $f''(x) > 0$, so x^* is a minimum point. However, if $f''(x) = 0$, as for $f(x) = x^3$ and $f(x) = x^4$ in $x^* = 0$, then higher-order derivatives should be considered to determine the status of x^*.

Theorem 3.3. *Let $f : \mathbb{R} \to \mathbb{R}$ be twice continuously differentiable in x^*. If $f'(x^*) = 0$ and $f''(x^*) > 0$, then x^* is a minimum point. If x^* is a minimum point, then $f'(x^*) = 0$ and $f''(x^*) \geq 0$.*

Extending Theorem 3.3 toward functions of several variables requires studying $\varphi_r''(0)$ in a stationary point x^*, where $\varphi_r(\lambda) = f(x^* + \lambda r)$. According to (3.10) we should know the sign of

$$\varphi_r''(0) = r^T H(x^*) r \qquad (3.30)$$

in all directions r. Expression (3.30) is a quadratic form. The derivation of Theorem 3.3 via (3.13) also applies for functions in several variables via (3.15).

Theorem 3.4. *Let $f : \mathbb{R}^n \to \mathbb{R}$ be twice continuously differentiable in x^*. If $\nabla f(x^*) = 0$ and $H(x^*)$ is positive definite, then x^* is a minimum point. If x^* is a minimum point, then $\nabla f(x^*) = 0$ and $H(x^*)$ is positive semidefinite.*

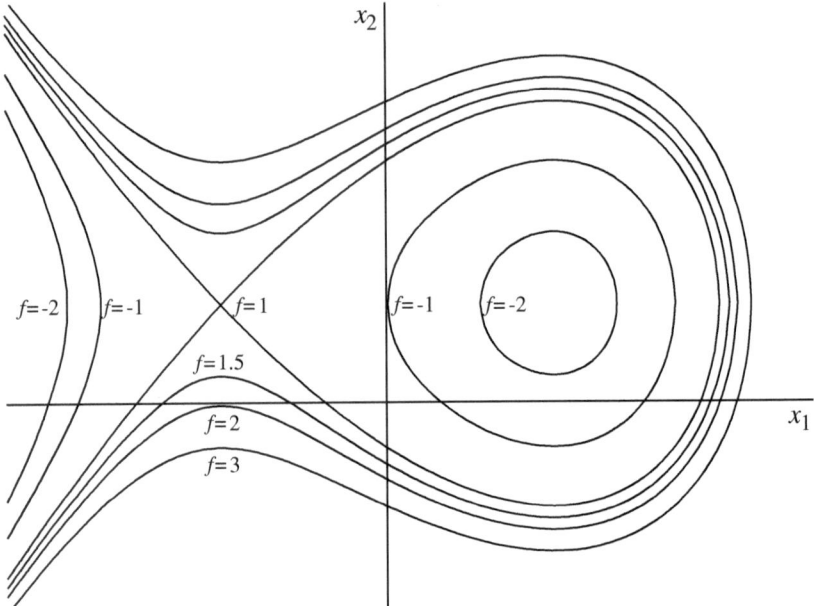

Fig. 3.8. Contours of $f(x) = x_1^3 - 3x_1 + x_2^2 - 2x_2$

Example 3.15. Consider the contours of $f(x) = x_1^3 - 3x_1 + x_2^2 - 2x_2$ in Figure 3.8. A minimum point and a saddle point can be recognized. Both are stationary points, but the Hessean has a different character. The gradient is $\nabla f(x) = \begin{pmatrix} 3x_1^2 - 3 \\ 2x_2 - 2 \end{pmatrix}$ and the Hessean $H(x) = \begin{pmatrix} 6x_1 & 0 \\ 0 & 2 \end{pmatrix}$. The eigenvalues of the Hessean are $6x_1$ and 2. The stationary points are determined by $\nabla f(x) = 0$; $x_1^* = (1,1)^T$ and $x_2^* = (-1,1)^T$.

$H(x_1^*) = \begin{pmatrix} 6 & 0 \\ 0 & 2 \end{pmatrix}$ is positive definite and $H_f(x_2^*) = \begin{pmatrix} -6 & 0 \\ 0 & 2 \end{pmatrix}$ is indefinite. This means that x_1^* is a minimum point and x_2^* is not a minimum point.

3.5 Optimality conditions, binding constraints

To check the optimality of a given x^*, it should be verified that f is non-decreasing from x^* in all feasible directions r. Mathematical theorems have been formulated to help the verification. If one does not carefully consider the underlying assumptions, the applications of such theorems may lead to incorrect conclusions about the status of x^*. With the aid of illustrative examples we try to make the reader aware of possible mistakes and the value of the assumptions.

Many names are connected to the mathematical statements with respect to optimality when there are binding constraints, in contrast to the theorems mentioned before. Well-known conditions are the so-called Karush–Kuhn–Tucker conditions (KKT conditions). We will first have a look at the historic perspective; see Kuhn (1991).

J.L. Lagrange studied the questions of optimization subject to equality constraints back in 1813. In 1939, W. Karush presented in his M.Sc. thesis, conditions that should be valid for an optimum point with equality constraints. Independently F. John presented some optimality conditions for a specific problem in 1948. Finally, the first-order conditions really became known after a presentation given by H.W. Kuhn and A.W. Tucker to a mathematical audience at a symposium in 1950. Their names were connected to the conditions that are nowadays known as the KKT conditions. Important notions are:

- Regularity conditions (constraint qualifications).
- Duality. We will sketch the relation with Linear Programming.
- Complementarity. This idea is important related to the distinction between binding and nonbinding constraints.

3.5.1 Lagrange multiplier method

The KKT conditions are often explained with the so-called Lagrange function or Lagrangean. It was developed for equality constraints $g_i(x) = 0$, but can also be applied to inequality constraints $g_i(x) \leq 0$.

$$L(x, u) = f(x) + \sum u_i g_i(x), \tag{3.31}$$

where $f(x)$ is the objective function that should be *minimized*. The constraints with respect to $g_i(x)$ are added to the objective function with so-called Lagrange multipliers u_i that can be interpreted as dual variables. The most important property of this function is that under some conditions it can be shown that for any minimum point x^* of (1.1), there exists a dual solution u^* such that (x^*, u^*) is a saddle point of $L(x, u)$ via

$$x^*, u^* \quad \text{is a solution of} \quad \min_x \max_u L(x, u). \tag{3.32}$$

So x^* is a minimum point of L (u^* constant) and u^* a maximum point. We are going to experiment with this idea. Why is it important to get some feeling for (3.32)? Often implicit use is made of (3.32) following the concept of the "Lagrange multiplier method." In this concept one uses the idea of the saddle point for putting the derivatives to u and x to zero and trying to find analytical solutions x^*, u^* of

$$\nabla L(x, u) = 0. \tag{3.33}$$

Example 3.16. In Example 3.1 (see Figure 3.1), one maximizes $U(x) = x_1 x_2$. The optimum has been determined graphically to be $x^* = (3, 3/2)$ where only

$g_1(x) = x_1 + 2x_2 - 6 \leq 0$ is a binding constraint. Given point x^* in the Lagrangean

$$L(x, u) = -U(x) + \sum u_i g_i(x)$$

one can put $u_2^* = u_3^* = 0$, because the second and third constraint $x_1 \geq 0$ and $x_2 \geq 0$ are nonbinding. The optimum point is the same if the second and third constraint are left out of the problem. This illustrates the notion of complementarity, that is also valid in Linear Programming; $u_i^* g_i(x^*) = 0$. If $u_2 = u_3 = 0$, then $L(x, u) = -x_1 x_2 + u_1(x_1 + 2x_2 - 6)$. So (3.33) leads to

$$\left. \begin{array}{l} \partial L/\partial x_1 = 0 \Rightarrow -x_2 + u_1 = 0 \\ \partial L/\partial x_2 = 0 \Rightarrow -x_1 + 2u_1 = 0 \\ \partial L/\partial u_1 = 0 \Rightarrow x_1 + 2x_2 - 6 = 0 \end{array} \right\} \Rightarrow$$

$$x_1^* = 3, x_2^* = 3/2, u_1^* = 3/2, U(x^*) = 4.5$$

is a unique solution and $x^* = (3, 3/2), u^* = (3/2, 0, 0)$ is a stationary point of the Lagrangean corresponding to the optimum. The value of $u_1^* = 3/2$ has the interpretation of shadow price; an additional (marginal) unit of budget results in $3/2$ units of additional utility. One can compare the values with the output of the Excel solver in the appendix.

The Lagrange multiplier method is slightly tricky:

1. Finding a stationary point analytically may not be easy.
2. An optimal solution may be one of (infinitely) many solutions of (3.33).
3. Due to some additional constraints, the saddle point (3.32) of L may not coincide with a solution of (3.33).
4. For the inequality constraints one should know in advance which $g_i(x) \leq 0$ are binding. Given a specific point x^* this is of course known.

These difficulties are illustrated by the following examples.

1. Finding a solution and 4. binding constraints

Analyzing a given point x^* in Example 3.16 is easy; (3.33) appears to be a linear set of equalities that can easily be solved. If the optimum point is not known, the binding constraints are unknown in (3.33). Furthermore, finding a solution for Example 3.16 is much harder when the objective function is changed to $U(x) = x_1 x_2^2$.

Example 3.17. Notice that finding a solution for Example 3.16 is not as easy as it seems. First of all we called $g_2(x) \leq 0$ and $g_3(x) \leq 0$ nonbinding constraints. If one by mistake puts $u_1 = 0$ ($g_1(x) \leq 0$ is nonbinding), the stationary point of the Lagrangean is $x^* = (0, 0), u^* = (0, 0, 0), g(x^*) = (6, 0, 0)$. This fulfills (3.33), but is neither an optimum point nor a solution of (3.32).

2. A solution of (3.33) is not an optimum point

Finding a solution is one difficulty. Another difficulty is that when a solution of (3.33) has been found, it does not necessarily correspond to a solution of the optimization problem.

Example 3.18. Consider the utility function of Figure 3.2, $U(x) = x_1^2 + x_2^2$. Following the same procedure as in Example 3.16 ($u_2 = u_3 = 0$) leads to

$$L(x, u) = -x_1^2 - x_2^2 + u_1(x_1 + 2x_2 - 6)$$
$$\nabla L(x, u) = 0 \quad \text{gives}$$
$$-2x_1 + u_1 = 0$$
$$-2x_2 + 2u_1 = 0$$
$$x_1 + 2x_2 = 6.$$

The solution of this system is $x_1^* = 6/5$, $x_2^* = 12/5$ with $U(x^*) = 7.2$ and $u_1^* = 12/5$, but not a maximum point of (3.1); over $x_1 + 2x_2 = 6$ it is even a minimum point.

Example 3.18 illustrates that the first-order conditions are necessary, but not sufficient. The optimum values for Example 3.18 can be found in the appendix.

3. The saddle point (3.32) is not a stationary point (3.33)

We focus on the case that an optimum point x^* corresponds to a saddle point of $L(x, u)$ but not to a stationary point; not all constraints are included in L. As shown before, the complementarity with respect to the inequalities should be taken into account. We illustrate this by considering a Linear Programming (LP) problem in the standard form

$$\max \{c^T x\}$$
$$Ax = b \tag{3.34}$$
$$x \geq 0.$$

The Lagrange function is formulated with respect to the equalities $Ax = b$ leaving the inequalities $x \geq 0$, where we do not know in advance which are binding and which nonbinding in the optimal solution. Given an optimal plan x^*, it is known which $x_j^* = 0$ and formulas exist to determine u^*, see Bazaraa et al. (1993). The Lagrangean of (3.34) is

$$L(x, u) = -c^T x + u^T (Ax - b). \tag{3.35}$$

Literature shows that a solution x^* of (3.34) is also a saddle point of (3.35), i.e.,

$$\min_{x \geq 0} \max_u L(x, u). \tag{3.36}$$

What does this mean? Notice that u is free and maximization results in an unbounded solution whenever $Ax \neq b$. Elaboration gives a logical result:

$$\min_{x \geq 0} \ [\max_u \{-c^T x + u^T (Ax - b)\}] = \min_{x \geq 0} \begin{cases} \infty \text{ if } Ax \neq b \\ -c^T x \text{ if } Ax = b \end{cases}$$

and also follows from $\partial L / \partial u_i = 0$. Setting the derivatives toward x_j to zero makes no sense, because we should know which x_j have a value of zero (basic versus nonbasic variables).

The Lagrange multiplier method via (3.32) and (3.33) can always be used to check the optimality of a given plan x^*, but is not always useful to find solutions.

It is noted for the interested reader that the dual problem (D) is defined by switching max and min in (3.36):

$$\max_u \ [\min_{x \geq 0} \ L(x, u)] =$$
$$\max_u \ [\min_{x \geq 0} \ \{(A^T u - c)^T x - b^T u\}] =$$
$$\max_u \ \begin{cases} -\infty \text{ if there exists an } i \text{ with } c_i > a_i^T u \\ -b^T u \text{ if } A^T u \leq c. \end{cases}$$

3.5.2 Karush–Kuhn–Tucker conditions

The Lagrange multiplier method may not always be appropriate for finding an optimum. On the other hand, an optimum point x^* (under regularity conditions and differentiability) should correspond to a stationary point of the Lagrangean (3.33) via the Karush–Kuhn–Tucker conditions, in which the notion of complementarity is more explicit.

Theorem 3.5. Karush–Kuhn–Tucker conditions
If x^* is a minimum point of (1.1), then there exist numbers u^* such that

$$-\nabla f(x^*) = \sum_i u_i^* \nabla g_i(x^*)$$
$$u_i^* g_i(x^*) = 0 \qquad\qquad complementarity$$
$$u_i^* \geq 0 \ for \ constraints \ g_i(x) \leq 0.$$

In mathematical terms this theorem shows us that the direction of optimization $(-\nabla f(x^*)$ in a minimization problem and $\nabla f(x^*)$ in a maximization problem) in the optimum is a combination of the gradients of the active constraints.

We first view this graphically and then go for an example. Point x^* is a minimum, if in any feasible direction r it is nondecreasing. A small positive step into a feasible direction cannot generate a lower objective function value. Graphically seen, the directions that point into the feasible area are related to the gradients of the active constraints (see Figure 3.9). Mathematically this can be seen as follows. If constraint $g_i(x) \leq 0$, is binding (active) in x^*, $g_i(x^*) = 0$ a direction r fulfilling $r^T \nabla g_i(x^*) < 0$ is pointing into the feasible area and a direction such that $r^T \nabla g_i(x^*) > 0$ points out of the area. In a minimum point x^* every feasible direction r should lead to an

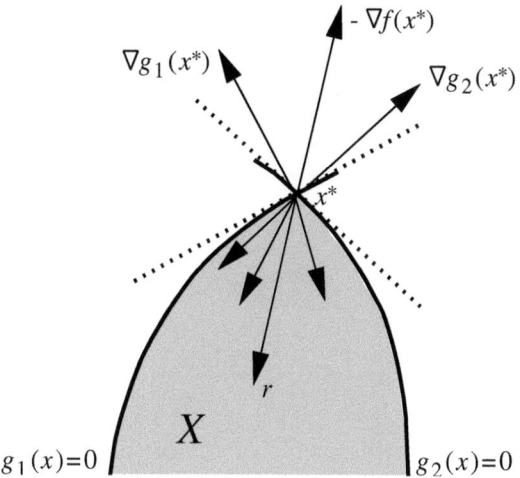

Fig. 3.9. Feasible directions

increase in the objective function value, i.e., $r^T\nabla f(x^*) \geq 0$. If a direction r fulfills $r^T\nabla g_i(x^*) < 0$ for every binding constraint, then it should also fulfill $r^T\nabla f(x^*) \geq 0$. Because of the KKT conditions $-\nabla f(x^*) = \sum u_i\nabla g_i(x^*)$ with $u_i \geq 0$ every feasible direction r fulfills

$$- r^T\nabla f(x^*) = \sum u_i r^T\nabla g_i(x^*) \geq 0. \tag{3.37}$$

So the KKT conditions are necessary to imply that x^* is a minimum point in all feasible directions. Graphically this means that arrow $-\nabla f(x^*)$ is situated in between the gradients $\nabla g_i(x^*)$ for all binding inequalities.

Example 3.19. Problem (3.1) with $U(x) = x_1^2 + x_2^2$ can be formulated as

$$\begin{aligned}
\min\{f(x) = & -x_1^2 - x_2^2\} \\
g_1(x) = & \ x_1 + 2x_2 - 6 \leq 0 \\
g_2(x) = & \quad\ -x_1 \qquad\ \leq 0 \\
g_3(x) = & \qquad\quad -x_2 \ \ \leq 0
\end{aligned}$$

so

$$\nabla f(x) = \begin{pmatrix} -2x_1 \\ -2x_2 \end{pmatrix}, \nabla g_1(x) = \begin{pmatrix} 1 \\ 2 \end{pmatrix}, \nabla g_2(x) = \begin{pmatrix} -1 \\ 0 \end{pmatrix}, \nabla g_3(x) = \begin{pmatrix} 0 \\ -1 \end{pmatrix}.$$

In the (local) minimum point $x_2^* = (0,3)^T$, g_1 and g_2 are binding and $g_3(x_2^*) = -3 < 0$ is nonbinding, so that $u_3^* = 0$.

$$-\nabla f(x_2^*) = \begin{pmatrix} 0 \\ 6 \end{pmatrix} = u_1^*\nabla g_1(x_2^*) + u_2^*\nabla g_2(x_2^*) + 0\nabla g_3(x_2^*) \Rightarrow$$

$$\begin{pmatrix} 0 \\ 6 \end{pmatrix} = u_1^*\begin{pmatrix} 1 \\ 2 \end{pmatrix} + u_2^*\begin{pmatrix} -1 \\ 0 \end{pmatrix} \Rightarrow u_1^* = 3, \ u_2^* = 3, \ u_3^* = 0.$$

For the global minimum point $x_1^* = (6,0)^T$ can be derived analogously:

$$-\nabla f \begin{pmatrix} 6 \\ 0 \end{pmatrix} = \begin{pmatrix} 12 \\ 0 \end{pmatrix} = u_1^* \begin{pmatrix} 1 \\ 2 \end{pmatrix} + 0 \begin{pmatrix} -1 \\ 0 \end{pmatrix} + u_3^* \begin{pmatrix} 0 \\ -1 \end{pmatrix} \Rightarrow$$

$$u_1^* = 12, \ u_2^* = 0, \ u_3^* = 24.$$

One can compare these values with the ones in the appendix.

Note that $x_3^* = \frac{1}{5} \begin{pmatrix} 6 \\ 12 \end{pmatrix}$ is a KKT point (not optimum) according to

$$-\nabla f(\frac{1}{5} \begin{pmatrix} 6 \\ 12 \end{pmatrix}) = \frac{1}{5} \begin{pmatrix} 12 \\ 24 \end{pmatrix} = \frac{12}{5} \nabla g_1 + 0 \nabla g_2 + 0 \nabla g_3.$$

Under regularity conditions, the KKT conditions are necessary for a point x^* to be optimum. The KKT conditions are not sufficient, as has been shown by Example 3.19. Similar to the case without binding constraints, *second-order conditions* exist based on the Hessean. Those conditions are far more complicated, because the sign of the second-order derivatives should be determined in the tangent planes of the binding constraints. We refer to the literature on the topic, such as Scales (1985), Gill et al. (1981) and Bazaraa et al. (1993). In the following section the notion of convexity will be discussed and its relation to the second-order conditions.

3.6 Convexity

Why deal with the mathematical notion of convexity? The relevance for a general NLP problem is mainly due to three properties. For a so-called convex optimization problem (1.1) applies:

1. If f and g_i are differentiable functions, a KKT point (and a stationary point) is also a minimum point. This means the KKT conditions are sufficient for optimality.
2. If a minimum point is found, it is also a global minimum point.
3. A maximum point can be found at the boundary of the feasible region. It is even a so-called extreme point.

Note that the notion of convexity is not directly related to differentiability. It is appropriate for property 1. The second and third property are also valid for nondifferentiable cases. How can one test the convexity of a specific problem? That is a difficult point. For many black-box applications and formulations in Chapter 2, where the calculation of the function is the result of a long calculation process, analysis of the formulas is not possible. The utility maximization examples in this chapter reveal their expressions and one can check the convexity. In economics literature where NLP is applied, other

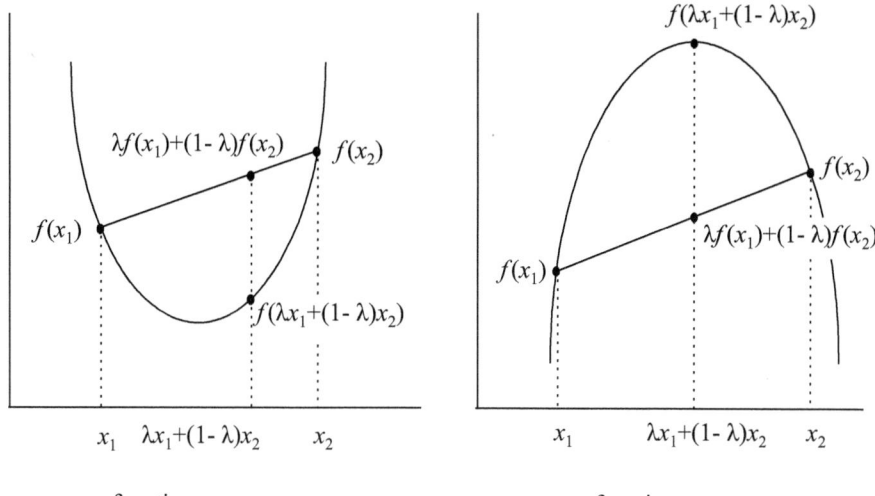

convex function   concave function

Fig. 3.10. Convex and concave functions

weaker assumptions can often be found; the functions g_i are quasi-convex. What is the meaning and the relation with the notion of convexity? This will be outlined. For a more detailed overview we refer to Bazaraa et al. (1993).

Definition 6. A function f is called *convex* when the chord between two points on the graph of f is nowhere below the graph, Figure 3.10. Mathematically:

$$f(\lambda x_1 + (1 - \lambda)x_2) \leq \lambda f(x_1) + (1 - \lambda)f(x_2) \quad 0 \leq \lambda \leq 1.$$

A function is *concave* if it is the other way around:

$$f(\lambda x_1 + (1 - \lambda)x_2) \geq \lambda f(x_1) + (1 - \lambda)f(x_2) \quad 0 \leq \lambda \leq 1.$$

In all other cases the terminology is that of nonconvex and nonconcave functions. In this definition some details are omitted; namely, f is defined on a so-called convex nonempty space. This is discussed later in Definition 7. It is not always easy in practice to show via the definition that a function is convex. Some examples are given.

Example 3.20. For a linear function $f(x) = c^T x$:

$$f(\lambda x_1 + (1 - \lambda)x_2) = c^T(\lambda x_1 + (1 - \lambda)x_2)$$
$$= \lambda c^T x_1 + (1 - \lambda)c^T x_2 = \lambda f(x_1) + (1 - \lambda)f(x_2).$$

By definition a linear function is as well convex as concave.

Example 3.21. For the quadratic function $f(x) = x^2$ the convexity question is given by

$$(\lambda x_1 + (1-\lambda)x_2)^2 \le \lambda x_1^2 + (1-\lambda)x_2^2 \quad \text{so}$$
$$\lambda x_1^2 + (1-\lambda)x_2^2 - (\lambda x_1 + (1-\lambda)x_2)^2 \ge 0 \ ?$$

Elaboration gives

$$\lambda x_1^2 + (1-\lambda)x_2^2 - \lambda^2 x_1^2 - (1-\lambda)^2 x_2^2 - 2\lambda(1-\lambda)x_1 x_2$$
$$= \lambda(1-\lambda)x_1^2 + \lambda(1-\lambda)x_2^2 - 2\lambda(1-\lambda)x_1 x_2 = \lambda(1-\lambda)(x_1 - x_2)^2 \ge 0$$
for $0 \le \lambda \le 1$.

Indeed $f(x) = x^2$ is convex.

3.6.1 First-order conditions are sufficient

We show that for a convex function f, a stationary point is a minimum point. This can be seen from the observation that a tangent line (plane) is below the graph of f; see Figure 3.11.

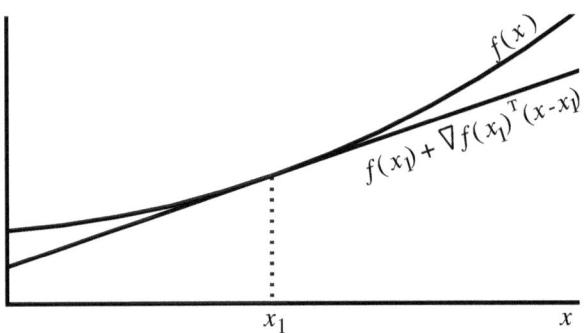

Fig. 3.11. Tangent plane below graph

Theorem 3.6. *Let f be a convex and continuously differentiable function on X. For any two points $x, x_1 \in X$*

$$f(x) \ge f(x_1) + \nabla f(x_1)^T (x - x_1). \tag{3.38}$$

This can be seen as follows. For a convex function f

$$f(\lambda x + (1-\lambda)x_1) \le \lambda f(x) + (1-\lambda)f(x_1).$$

So, $f(x_1 + \lambda(x - x_1)) \le f(x_1) + \lambda(f(x) - f(x_1))$; this means $f(x) - f(x_1) \ge \frac{f(x_1 + \lambda(x-x_1)) - f(x_1)}{\lambda}$. Limit $\lambda \to 0$ results at the right-hand side in the directional derivative f in x_1 in the direction $(x - x_1)$, so that $f(x) - f(x_1) \ge \nabla f(x_1)^T (x - x_1)$.

Now it follows directly from (3.38) that x^* is a minimum point, as in a stationary point x^*, $\nabla f(x^*) = 0$.

Theorem 3.7. *If f is convex in an ϵ-environment of stationary point x^*, then x^* is a minimum point of f.*

Convexity and the Hessean is positive semidefinite
Combining Theorem 3.7 with the second-order conditions of Theorem 3.4 shows a relationship between convexity and the Hessean for twice-differentiable functions.

Theorem 3.8. *Let $f : X \to \mathbb{R}$ be twice continuously differentiable on open set X: f is convex $\Leftrightarrow H_f$ is positive semidefinite on X.*

Theorem 3.8 follows from combining (3.27) and (3.38). The theorem shows that in some cases convexity can be checked.

Example 3.22. The function $f(x) = x_1^2 + 2x_2^2$ is convex.
The Hessean is $H_f = \begin{pmatrix} 2 & 0 \\ 0 & 4 \end{pmatrix}$. The eigenvalues of the Hessean are 2 and 4, so H_f is positive definite. Theorem 3.8 tells us that f is convex.

3.6.2 Local minimum point is global minimum point

For the notion of convex optimization, the definition of convex set is required. A *convex optimization problem* is defined as a problem where the objective function f is convex in case of minimization (concave in case of maximization) and feasible set X is a convex set.

Definition 7. Set X is called convex if for any pair of points $p, q \in X$ the chord between those points is also in X : $\lambda p + (1 - \lambda)q \in X$ for $0 \leq \lambda \leq 1$.

When is the feasible area X convex? In problem (1.1), X is defined by inequalities $g_i(x) \leq 0$ and equalities $g_i(x) = 0$. Linear equalities (LP) lead to a convex area, but if an equality $g_i(x) = 0$ is nonlinear, e.g., $x_1^2 + x_2^2 - 4 = 0$, a

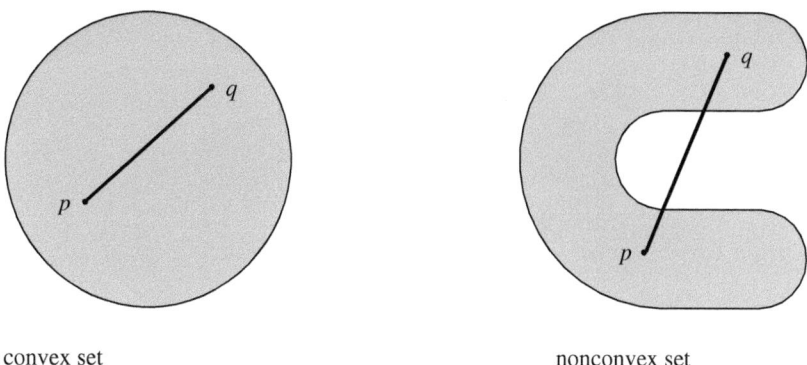

convex set   nonconvex set

Fig. 3.12. A convex and a nonconvex set

nonconvex area appears. In contrast to the mentioned equality, the inequality $x_1^2 + x_2^2 - 4 \le 0$ describes a circle with its interior and this is a convex set. Considering the inequality $g_i(x) \le 0$ more abstractly, it is a level set of the function $g_i(x)$. The relation with convex functions is given in Theorem 3.9.

Theorem 3.9. *Let* $g : X \to \mathbb{R}$ *be a convex function on a convex set* X *and* $h \in \mathbb{R}$. *Level set* $S_h = \{x \in X \mid g(x) \le h\}$ *is a convex set.*

The proof proceeds as follows. Given two points $x_1, x_2 \in S_h$ so $g(x_1) \le h$ and $g(x_2) \le h$. The convexity of g shows that point $x = \lambda x_1 + (1 - \lambda)x_2$ in between x_1 and x_2 is also in S_h:

$$g(x) = g(\lambda x_1 + (1-\lambda)x_2) \le \lambda g(x_1) + (1-\lambda)g(x_2) \le \lambda h + (1-\lambda)h = h. \quad (3.39)$$

A last property often mentioned in the literature is that the functions g_i are *quasi-convex*. This is a weaker assumption than convexity for which Theorem 3.9 also applies. To be complete, the definition is given here. The reader can derive the variant of inequality (3.39).

Definition 8. A function $f : X \to \mathbb{R}$ on a nonempty convex set X is called quasi-convex if for any pair $x_1, x_2 \in X$,

$$f(\lambda x_1 + (1 - \lambda)x_2) \le \text{maximum } \{f(x_1), f(x_2)\} \quad 0 \le \lambda \le 1.$$

The notion of a convex optimization problem (1.1) is important for the three properties we started with. The KKT conditions are sufficient to determine the optimality of a stationary point in a convex optimization problem, see Bazaraa et al. (1993). Property 2 (local is global) can now be derived.

Theorem 3.10. *Let* f *be convex on a convex set* X, *then every local minimum point is a global minimum point.*

Showing the validity of Theorem 3.10 is usually done in the typical mathematical way of demonstrating that assuming nonvalidity will lead to a contradiction. For a local minimum point x^*, an ϵ-environment W of x^* exists where x^* is minimum; $f(x) \ge f(x^*)$, $x \in X \cap W$. Suppose that Theorem 3.10 is not

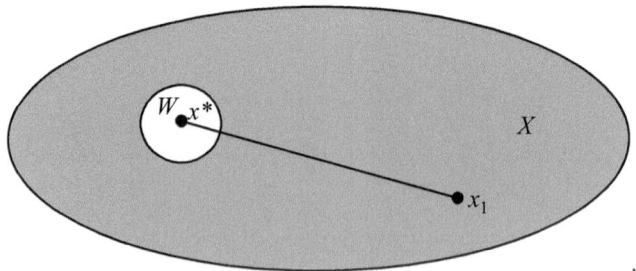

Fig. 3.13. Local, nonglobal does not exist

true. Then a point $x_1 \in W$ should exist such that $f(x_1) < f(x^*)$. By logical steps and the convexity of f and X it can be shown that the existence of x_1 leads to a contradiction. Points on the line between x_1 and x^* are situated in X, $x \in X$ and can be described by $x = \lambda x_1 + (1 - \lambda)x^*$, $0 \le \lambda \le 1$. Convexity of f implies

$$f(x) = f(\lambda x_1 + (1 - \lambda)x^*) \le \lambda f(x_1) + (1 - \lambda)f(x^*) < \lambda f(x^*) + (1 - \lambda)f(x^*) = f(x^*). \tag{3.40}$$

So the convexity of f and the assumption $f(x_1) < f(x^*)$ implies that all points on the chord between x_1 and x^* have an objective value lower than $f(x^*)$. For λ small, the point x is situated in W in contradiction to x^* being a local minimum. So the assumption that a point x_1 exists with $f(x_1) < f(x^*)$ cannot be true.

The practical importance of Theorem 3.10 is that software like GINO, GAMS/ MINOS, and the Excel solver return a local minimum point depending on the starting value. If one wants to be certain it is a global minimum point, then the optimization problem should be analyzed further on convexity. We already have seen that this may not be easy.

3.6.3 Maximum point at the boundary of the feasible area

The last mentioned property is: for a convex function a maximum point (if it exists) can be found at the boundary of the feasible area. A special case of this property is Linear Programming.

Theorem 3.11. *Let* $f : X \to \mathbb{R}$ *be a convex function on a closed set* X. *If* f *has a maximum on* X, *then there exists a maximum point* x^* *that is an extreme point.*

Mathematically extreme means that x^* cannot be written as a convex combination of two other points in X. A typical extreme point is a vertex (corner point). At the boundary of a circle, all points are extreme points. The proof of Theorem 3.11 also uses contradiction. The proof is constructed by assuming that there is an interior maximum point x^* ; more exactly, a maximum point x^* with a higher function value than the points at the boundary. Point x^* can be written as a convex combination of two points x_1 and x_2 at the boundary:

$$f(x^*) > f(x_1), \; f(x^*) > f(x_2) \text{ and } x^* = \lambda x_1 + (1 - \lambda)x_2.$$

Just like in (3.40) this leads to a contradiction:

$$f(x^*) \le \lambda f(x_1) + (1 - \lambda)f(x_2) < \lambda f(x^*) + (1 - \lambda)f(x^*) = f(x^*).$$

The consequence in this case is that if the feasible area is a polytope, one can limit the search for a maximum point to the vertices of the feasible area. Life does not necessarily become very easy with this observation; the number of vertices can explode in the number of decision variables. A traditional example showing this and also giving a relation between NLP and combinatorial optimization is the following.

Example 3.23.
$$\max \{f(x) = \sum(x_i - \varepsilon)^2\}$$
$$-1 \le x_i \le 1, \quad i = 1, \dots, n,$$

where ε is a small number, e.g., 0.01. The problem describes the maximization of the distance to a point that is nearly in the middle of a square/cube. With increasing dimension n, the number of vertices explodes, increases with 2^n. Every vertex is a local maximum point. Moreover, this problem has a multiple of KKT points that are not maximum points. For the case of a cube $(n = 3)$ for instance 18.

3.7 Summary and discussion points

- Optimum points of an NLP problem may be found in the interior, on the boundary of the feasible set or in extreme points.
- Trade-off curves giving the optimal solution are typically nonlinear in changing parameter values. Nondifferentiable points also occur due to constraints changing their status being binding.
- First-order conditions are typically based on the concept of stationary points and Karush–Kuhn–Tucker points.
- Second-order conditions are based on the status of the Hessean on being positive definite.
- Convexity of a problem gives that first-order conditions are sufficient for points to be minimum points.
- For convex optimization problems, local minimum points are also global minimum points.
- A convex objective function finds its maxima at the extreme points of the feasible set.

3.8 Exercises

1. Solve the following NLP problem graphically:

$$\min(x_1 - 3)^2 + (x_2 - 2)^2$$

subject to the constraints

$$x_1^2 - x_2 - 3 \leq 0$$
$$x_2 - 1 \leq 0$$
$$-x_1 \leq 0.$$

2. Designing a desk with length x and y wide, the following aspects appear:
   - The surface has to be as big as possible: max xy.
   - The costs of the expensive edge should not be too large: $2x + 2y \leq 8$.
   - The desk should not be too wide: $y \leq b$.

   Solve this NLP problem graphically for $b = 1$.

   What happens to the optimal surface when b increases?

3. In Example 3.4 determine V, x_1^* and the (E, V)-curve when there is negative correlation according to $\sigma_{12} = -\frac{1}{2}$.

4. Is the function $f(x) = \sqrt{(x_1^2 + x_2^2)}$ differentiable in 0?

5. Determine gradient and Hessean of $f(x) = x_1^2 x_2^2 e^{x_3}$.

6. Derive the second-order Taylor approximation of $f(x) = x_1 e^{x_2}$ around 0.

7. Given $f(x) = 2x_1^2 + x_2^4$. Derive and draw the contours corresponding to a function value of 3, of $f(x)$ and the first- and second-order Taylor approximation around $(1, 1)^T$.

8. Let $f(x) = 4x_1 x_2 + 6x_1^2 + 3x_2^2$. Write $f(x)$ as a quadratic function (3.16). Determine stationary point and eigenvalues and eigenvectors of A.

9. Let $f(x) = -1 - 2x_1 - x_2 + x_1 x_2 + x_1^2$. Write $f(x)$ as a quadratic function (3.16). Determine stationary point and eigenvalues and eigenvectors of A.

10. Let $f(x) = (x_1 - x_2)^2$. Write $f(x)$ as a quadratic function (3.16). Determine the stationary points and eigenvalues and eigenvectors of A. Are the stationary points minimum points?

11. Determine the minima of
    (a) $f(x) = x_1^2 + x_2^2$
    (b) $f(x) = \sqrt{x_1^2 + x_2^2}$
    (c) $f(x) = x_1 x_2$

12. Given function $f(x) = x_1^3 - x_2^3 - 6x_1 + x_2$:
    (a) Determine gradient and Hessean of $f(x)$.
    (b) Determine the stationary points of $f(x)$.
    (c) Which point is a minimum point and which are saddle points?

13. Given utility function $U(x) = x_1^2 x_2$ and budget constraint $x_1 + x_2 = 3$. Determine the stationary point of the Lagrangean maximizing the utility function subject to the budget constraint.

14. Given NLP problem

$$\min (x_1 - 3)^2 + (x_2 - 2)^2$$

subject to

$$x_1^2 + x_2^2 \leq 5$$
$$x_1 + 2x_2 \leq 4$$
$$-x_1 \qquad \leq 0$$
$$-x_2 \leq 0.$$

(a) Determine graphically the optimal solution x^*.
(b) Check the Karush–Kuhn–Tucker (KKT) conditions for x^*.
(c) What happens when the right-hand side of the second constraint (4) increases?

15. Given LP problem

$$\max \ 2x_1 + 2x_2 \qquad \text{(P)}$$
$$x_1 + 2x_2 \leq b$$
$$4x_1 + 2x_2 \leq 10$$
$$x_1, x_2 \geq 0.$$

(a) Solve (P) for $b = 4$ with the simplex method.
(b) Check the KKT conditions in the optimum point.
(c) Compare the values of the KKT multipliers to the solution of the dual problem.
(d) What happens to the optimum if b increases?

16. Given the concave optimization problem P:

$$\min \ -(x_1 - 1)^2 - x_2^2 \qquad \text{(P)}$$
$$2x_1 + x_2 \leq 4$$
$$x_1, x_2 \geq 0.$$

(a) Determine graphically the local and global minimum points of P.
(b) Show that the minimum points fulfill the KKT conditions.
(c) Point $(0,0)$ fulfills the KKT conditions. Show via the definition that $(0,0)^T$ is not a local minimum point.
(d) Give another point that fulfills the KKT conditions, but is not a minimum point.

17. Show $f(x) = \max\{g_1(x), g_2(x)\}$ is convex if $g_1(x)$ and $g_2(x)$ are convex.

18. Given a convex continuous function $f : \mathbb{R}^n \to \mathbb{R}$. Show that its epigraph $\{(x, \alpha) \in \mathbb{R}^{n+1} | \alpha \geq f(x)\}$ is a convex set.

19. Check whether $f(x) = 2x_1 + 6x_2 - 2x_1^2 - 3x_2^2 + 4x_1x_2$ is convex.

20. Determine the validity of Theorem 3.6 for $f : (0, \infty) \to \mathbb{R}$ with $f(x) = 1/x$.

21. Given a convex optimization problem, i.e., the objective function f is convex as well as the feasible region X.
(a) Can the problem have more than one minimum?
(b) Can the problem have more than one minimum point?
(c) Can the problem have exactly two global minimum points?

22. Given quadratic function $f(x) = 2x_1^2 + x_2^2 - 2x_1x_2 - 6x_1 + 1$ and the feasible area X given by $3 \le x_1 \le 6$ and $0 \le x_2 \le 6$.
    (a) Show that f is convex.
    (b) Can f have more than one minimum on X?
    (c) Determine the minima of f on X.
    (d) Determine all maximum points of f on X via the KKT conditions.

23. Given problem

$$\min_X f(x) = x_1x_2, \quad X = \{x \in \mathbb{R}^2 | 2x_1 + x_2 \ge 6, x_1 \ge 1, x_2 \ge 1\}. \quad (3.41)$$

    (a) Determine graphically all minimum points of (3.41).
    (b) Show that the minimum points fulfill the KKT conditions.
    (c) Show feasible point $x = (3, 1)^T$ does not fulfill the KKT conditions.
    (d) Give a point that fulfills KKT conditions, but is not a minimum point.

24. Given $f(x) = 24x_1 + 14x_2 + x_1x_2$ and point $x_0 = (2, 10)^T$ with $f(x_0) = 208$.
    (a) Determine gradient and Hessean of f.
    (b) Give a descent direction r in x_0.
    (c) Is f convex in direction r?

3.9 Appendix: Solvers for Examples 3.2 and 3.3

Input and output of GINO for Example 3.3

```
MODEL:
     1) MAX= X1 ^ 2 + X2 ^ 2 ;
     2) X1 + 2 * X2 < 6 ;
     3) X1 > 0 ;
     4) X2 > 0 ;
   END


   SOLUTION STATUS:  OPTIMAL TO TOLERANCES.  DUAL
   CONDITIONS:  SATISFIED.

            OBJECTIVE FUNCTION VALUE

     1)        36.000000

   VARIABLE         VALUE        REDUCED COST
        X1        6.000000          .000000
        X2         .000000          .000000

      ROW   SLACK OR SURPLUS          PRICE
       2)         .000000        12.000010
       3)        6.000000          .000000
       4)         .000000       -24.000009
```

Fig. 3.14. Input and output GINO

The optimal values for the variables, Lagrange multiplier and the status of the constraints in the optimal solution can be recognized.

Input and output of GAMS/MINOS for Example 3.3

```
Variables
X1
X2
NUT;

POSITIVE VARIABLES X1,X2;

EQUATIONS

BUDGET
NUTD;

BUDGET.. X1+2*X2=L=6;
NUTD.. NUT=E=X1*X1+X2*X2;

MODEL VBNLP /ALL/

SOLVE VBNLP USING NLP MAXIMIZING NUT
```

Somewhere in the OUTPUT (7 pages) the optimal values of the variables, constraints and shadow prices can be found. Here MINOS finds the global optimum. For another starting value it may find the local optimum.

```
GAMS 2.25.081 386/486 G e n e r a l    A l g e b r a i c    M o d e l i n g    S y s t e m
**** SOLVER STATUS     1 NORMAL COMPLETION
**** MODEL STATUS      2 LOCALLY OPTIMAL
**** OBJECTIVE VALUE          36.0000

    RESOURCE USAGE, LIMIT          0.220      1000.000
    ITERATION COUNT, LIMIT         0          1000
    EVALUATION ERRORS              0          0

        M I N O S    5.3   (Nov 1990)        Ver: 225-386-02
        = = = = =
        B. A. Murtagh, University of New South Wales
          and
        P. E. Gill,  W. Murray,  M. A. Saunders and M. H. Wright
        Systems Optimization Laboratory, Stanford University.

    EXIT -- OPTIMAL SOLUTION FOUND
    MAJOR ITNS, LIMIT              1      200
    FUNOBJ, FUNCON CALLS           4        0
    SUPERBASICS                    0
    INTERPRETER USAGE           0.00
    NORM RG / NORM PI      0.000E+00

                        LOWER      LEVEL      UPPER     MARGINAL

    ---- EQU BUDGET      -INF      6.000      6.000      12.000
    ---- EQU NUTD          .         .          .        -1.000

                        LOWER      LEVEL      UPPER     MARGINAL

    ---- VAR X1            .       6.000      +INF         .
    ---- VAR X2            .         .        +INF      -24.000
    ---- VAR NUT         -INF     36.000      +INF         .
```

Fig. 3.15. Part of output GAMS

Output of Excel solver for Example 3.2

**Microsoft Excel 8.0e Answer
Report
Worksheet:
[xlsolver.xls]Sheet1**

Target Cell
(Max)

Cell	Name	Original Value	Final Value
D9	x1*x2	0	4.5

Adjustable Cells

Cell	Name	Original Value	Final Value
E6	x1	6	3
F6	x2	0	1.5

Constraints

Cell	Name	Cell Value	Formula	Status	Slack
G6	x1+2*x2	6	G6<=6	Binding	0
E6	x1	3	E6>=0	Not Binding	3
F6	x2	1.5	F6>=0	Not Binding	1.5

**Microsoft Excel 8.0e Sensitivity Report
Worksheet: [xlsolver.xls]Sheet1**

Adjustable Cells

Cell	Name	Final Value	Reduced Gradient
E6	x1	3	0
F6	x2	1.5	0

Constraints

Cell	Name	Final Value	Lagrange Multiplier
G6	x1+2*x2	6	1.5

Fig. 3.16. Output Excel solver for Example 3.2

4

Goodness of optimization algorithms

4.1 Effectiveness and efficiency of algorithms

In this chapter, several criteria are discussed to measure the effectiveness and efficiency of algorithms. Moreover, examples of basic algorithms are analyzed. Global Optimization (GO) concepts such as *region of attraction, level set, probability of success* and *performance graph* are introduced. To investigate optimization algorithms, we should say what we mean by them in this book; an algorithm is a description of steps, preferably implemented into a computer program, which finds an approximation of an optimum point. The aims can be several: reach a local optimum point, reach a global optimum point, find all global optimum points, reach all global and local optimum points. In general, an algorithm generates a series of points x_k that approximate an optimum point. According to the generic description of Törn and Žilinskas (1989):

$$x_{k+1} = Alg(x_k, x_{k-1}, \ldots, x_0, \xi), \tag{4.1}$$

where ξ is a random variable and index k is the iteration counter. This represents the idea that a next point x_{k+1} is generated based on information in all former points $x_k, x_{k-1}, \ldots, x_0$ (x_0 usually being the starting point) and possibly a random effect. This leads to three classes of algorithms discussed here:

- Nonlinear optimization algorithms, that from a starting point try to reach the "nearest" local minimum point. These are described in Chapter 5.
- Deterministic GO methods which guarantee to approach the global optimum and require a certain mathematical structure. Attention is paid in Chapter 6, but also several heuristics are discussed.
- Stochastic GO methods based on the random generation of feasible trial points and nonlinear local optimization procedures. Those are discussed in Chapter 7.

E.M.T. Hendrix and B.G.-Tóth, *Introduction to Nonlinear and Global Optimization*, Springer Optimization and Its Applications 37, DOI 10.1007/978-0-387-88670-1_4, © Springer Science+Business Media, LLC 2010

We will consider several examples illustrating two questions to be addressed to investigate the quality of algorithms (see Baritompa and Hendrix, 2005).

- Effectiveness: does the algorithm find what we want?
- Efficiency: what are the computational costs?

Several measurable performance indicators can be defined for these criteria.

4.1.1 Effectiveness

Consider minimization algorithms. Focusing on effectiveness, there are several targets a user may have:

1. To discover all global minimum points. This of course can only be realized when the number of global minimum points is finite.
2. To detect at least one global optimum point.
3. To find a solution with a function value as low as possible.
4. To produce a uniform covering of a near-optimal or success region. This idea as introduced by Hendrix and Klepper (2000) can be relevant for population-based algorithms.

The first and second targets are typical satisfaction targets; was the search successful or not? What are good *measures of success*? In the literature, convergence is often used, i.e., $x_k \to x^*$, where x^* is one of the minimum points. Alternatively one observes $f(x_k) \to f(x^*)$. In tests and analyses, to make results comparable, one should be explicit in the definitions of success. We need not only specify ϵ and/or δ such that

$$\|x_k - x^*\| < \epsilon \text{ and/or } f(x_k) < f(x^*) + \delta \tag{4.2}$$

but also specify whether success means that there is an index K such that (4.2) is true for all $k > K$. Alternatively, success may mean that a record $\min_k f(x_k)$ has reached level $f(x^*) + \delta$. Whether the algorithm is effective also depends on its stochastic nature. When we are dealing with stochastic algorithms, effectiveness can be expressed as the probability that a success has been reached. In analysis, this probability can be derived from sufficient

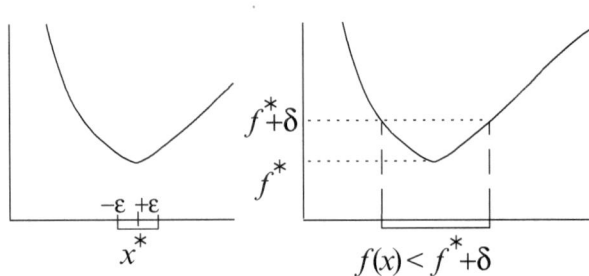

Fig. 4.1. Success region based on ϵ environment or $f^* + \delta$ level set

assumptions on the behavior of the algorithm. In numerical experiments, it can be estimated by counting repeated runs how many times the algorithm converges. We will give some examples of such analysis. In Section 4.2.4 we return to the topic of efficiency and effectiveness considered simultaneously.

4.1.2 Efficiency

Globally efficiency is defined as the effort the algorithm needs to be successful. A usual indicator for algorithms is the (expected) number of *function evaluations* necessary to reach the optimum. This indicator depends on many factors such as the shape of the test function and the termination criteria used. The indicator more or less suggests that the calculation of function evaluations dominates the other computations of the algorithm. Several other indicators appear in the literature.

In nonlinear programming (e.g., Scales, 1985; Gill et al., 1981) the concept of *convergence speed* is common. It deals with the convergence limit of the series x_k. Let $x_0, x_1, \ldots, x_k, \ldots$ converge to point x^*. The largest number α for which

$$\lim_{k \to \infty} \frac{\|x_{k+1} - x^*\|}{\|x_k - x^*\|^\alpha} = \beta < \infty \tag{4.3}$$

gives the order of convergence, whereas β is called the convergence factor. In this terminology, the special instances are:

- linear convergence with $\alpha = 1$ and $\beta < 1$
- quadratic convergence with $\alpha = 2$ and $0 < \beta < 1$
- superlinear convergence: $1 < \alpha < 2$ and $\beta < 1$, i.e., $\beta = 0$ if $\alpha = 1$ in (4.3).

Mainly in deterministic GO algorithms, information on past evaluations is stored in the computer memory. This requires efficient data handling for looking up necessary information during the iterations. Furthermore, *memory requirements* become a part of the computational burden as retrieving actions cannot be neglected compared to the computational effort due to function evaluations.

In stochastic GO algorithms, an efficiency indicator is the *success rate* defined as the probability that the next iterate is an improvement on the record value found thus far, i.e., $P(f(x_k) < \min_{l=1,\ldots,k-1} f(x_l))$. Its theoretical relevance to convergence speed was analyzed by Zabinsky and Smith (1992) and Baritompa et al. (1995), who showed that a fixed success rate of an effective algorithm (in the sense of so-called uniform covering; see, e.g., Hendrix and Klepper, 2000) gives an algorithm with the expected number of function evaluations growing polynomially with the dimension of the problem. However, empirical measurements can only be established in the limit when such an algorithm stabilizes, and only for specifically designed test cases (Hendrix et al., 2001).

We do not go deeper into theoretical aspects of performance indicators here. Instead some basic algorithms are introduced and analyzed. In Section 4.3, systematic investigation of algorithms is expanded upon.

4.2 Some basic algorithms and their goodness

4.2.1 Introduction

In this section, several classes of algorithms are analyzed for effectiveness and efficiency. Two test cases are introduced first for which the performance of the algorithms are investigated. We consider the minimization of

$$g(x) - \sin(x) + \sin(3x) + \ln(x), \ x \in [3, 7]. \tag{4.4}$$

Function g is depicted in Figure 4.2 and has three minimum points on the interval, at $x^* = 3.73$, $x^* = 5.65$ and $x^* = 7$. The global minimum is attained at $x^* = 3.73$, where $g(x^*) = -0.220$. The derivative function is

$$g'(x) = \cos(x) + 3\cos(3x) + \frac{1}{x} \tag{4.5}$$

on the interval $[3, 7]$. Alternatively to function g, we introduce a function h with more local minimum points by adding to function g a bubble function based on $\mathrm{frac}(x) = x - \mathrm{round}(x)$ where $\mathrm{round}(x)$ rounds x to the nearest integer. Now the second case is defined as

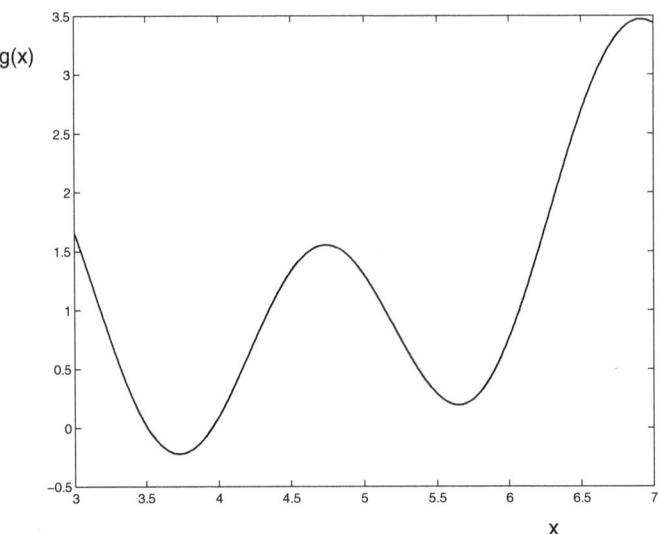

Fig. 4.2. Test case $g(x)$ with three optima

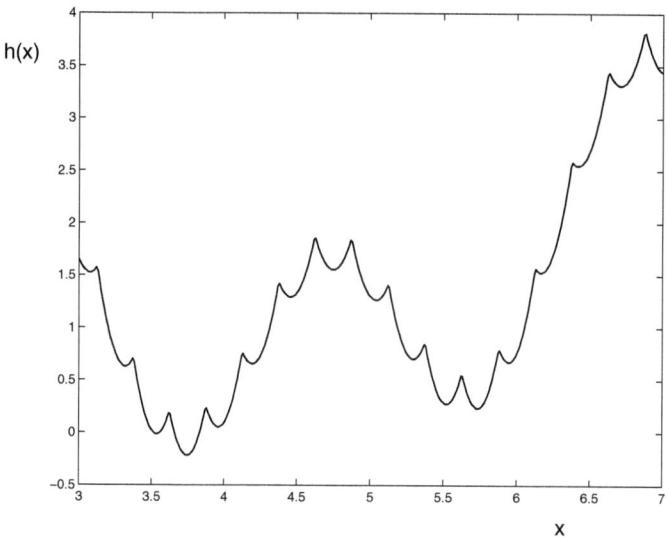

Fig. 4.3. Test case $h(x)$ with 17 optima

$$h(x) = g(x) + 1.5\text{frac}^2(4x). \tag{4.6}$$

In Figure 4.3, the graph of function h is shown. It has 17 local minimum points on the interval $[3, 7]$. Although neither g nor h is convex on the interval, at least function h is piecewise convex on the intervals in between the points of $\mathbb{S} = \{x = \frac{1}{4}k + \frac{1}{8}, k \in \mathbb{Z}\}$. At these points, h is not differentiable. For the rest of the interval one can define the derivative

$$h'(x) = g'(x) + 12 \times \text{frac}(4x) \quad \text{for} \quad x \notin \mathbb{S}. \tag{4.7}$$

The global minimum point of h on $[3, 7]$ is shifted slightly compared to g toward $x^* = 3.75$, where $h(x^*) = -0.217$.

In the following sections, we will test algorithms on their ability to find minima of these two functions. One should set a target on what is considered an acceptable or successful result. For instance, one can aim at detecting a local minimum or detecting the global minimum. For the neighborhood we will take an acceptance of $\epsilon = 0.01$. For determining an acceptable low value of the objective function we take $\delta = 0.01$. Notice that ϵ represents 0.25% of the argument range $[3, 7]$ and δ is about 0.25% of the function values range.

4.2.2 NLP local optimization: Bisection and Newton

Two nonlinear programming algorithms are sketched and their performance measured for the two test cases. Both are further elaborated in Chapter 5. First the bisection algorithm is considered.

Algorithm 1 Bisect($[l, r], f, \epsilon$)

Set $k := 0$, $l_0 = l$ and $r_0 = r$
while ($r_k - l_k > \epsilon$)
$\quad x_k := \frac{l_k + r_k}{2}$
$\quad$**if** ($f'(x_k) < 0$)
$\quad\quad l_{k+1} := x_k$ and $r_{k+1} := r_k$
$\quad$**else**
$\quad\quad l_{k+1} := l_k$ and $r_{k+1} := x_k$
$\quad k := k + 1$
endwhile

The algorithm departs from a starting interval $[l, r]$ that is halved iteratively based on the sign of the derivative in the midpoint. This means that the method is only applicable when the derivative is available at the generated midpoints. The point x_k converges to a minimum point within the interval $[l, r]$. If the interval contains only one minimum point, it converges to that. In our test cases, several minima exist and one can observe the convergence to one of them. The algorithm is effective in the sense of converging to a local (nonglobal) minimum point for both cases. Another starting interval could have led to another minimum point. In the end, we are certain that the current iterate x_k is not further away than ϵ from a minimum point. Alternative stopping criteria like convergence of function values or derivatives going to zero are possible for this algorithm. The current stopping criterion is easy for analysis of efficiency. One question could be: How many iterations (i.e., cor-

Table 4.1. Bisection for functions g and h, first 6 iterations

k	l_k	r_k	x_k	$g'(x_k)$	$g(x_k)$	l_k	r_k	x_k	$h'(x_k)$	$h(x_k)$
		function g						function h		
0	3.00	7.00	5.00	-1.80	1.30	3.00	7.00	5.00	-1.80	1.30
1	5.00	7.00	6.00	3.11	0.76	5.00	7.00	6.00	3.11	0.76
2	5.00	6.00	5.50	-1.22	0.29	5.00	6.00	5.50	-1.22	0.29
3	5.50	6.00	5.75	0.95	0.24	5.50	6.00	5.75	0.95	0.24
4	5.50	5.75	5.63	-0.21	0.20	5.50	5.75	5.63	-6.21	0.57
5	5.63	5.75	5.69	0.36	0.20	5.63	5.75	5.69	-2.64	0.29
6	5.63	5.69	5.66	0.07	0.19	5.69	5.75	5.72	-0.85	0.24

responding derivative function evaluations) are necessary to come closer than ϵ to a minimum point? The bisection algorithm is a typical case of linear convergence with a convergence factor of $\frac{1}{2}$, $\frac{|r_{k+1} - l_{k+1}|}{|r_k - l_k|} = \frac{1}{2}$. This means one can determine the number of iterations necessary for reaching ϵ-convergence:

$$| r_k - l_k | = (\tfrac{1}{2})^k \cdot | r_0 - l_0 | < \epsilon \quad \Rightarrow$$
$$(\tfrac{1}{2})^k < \frac{\epsilon}{| r_0 - l_0 |} \quad \Rightarrow \quad k > \frac{\ln \epsilon - \ln | r_0 - l_0 |}{\ln \frac{1}{2}}.$$

Algorithm 2 Newt($[l, r], x_0, f, \alpha$)

Set $k := 0$,
while $(\mid f'(x_k) \mid > \alpha)$

$\quad x_{k+1} := x_k - \frac{f'(x_k)}{f''(x_k)}$

$\quad\quad\quad$! safeguard for staying in interval

$\quad$ **if** $(x_{k+1} < l)$, $x_{k+1} := l$

$\quad$ **if** $(x_{k+1} > r)$, $x_{k+1} := r$

$\quad$ **if** $(x_{k+1} = x_k)$, STOP

$\quad k := k + 1$

endwhile

The example case requires at least 9 iterations to reach an accuracy of $\epsilon = 0.01$.

An alternative for finding the zero point of an equation, in our case the derivative, is the so-called *method of Newton*. The idea is that its efficiency is known to be superlinear (e.g., Scales, 1985), so it should be faster than bisection. We analyze its efficiency and effectiveness in the two test cases.

In general, the aim of the Newton algorithm is to converge to a point where the derivative is zero. Depending on the starting point x_0, the method may converge to a minimum or maximum. Also, it may not converge at all, for instance when a minimum point does not exist. Specifically in the version of Algorithm 2, a safeguard is built in to ensure the iterates remain in the interval; it can converge to a boundary point. If x_0 is in the neighborhood of a minimum point where f is convex, then convergence is guaranteed and the algorithm is effective in the sense of reaching a minimum point. Let us consider what happens for the two test cases.

When choosing the starting point x_0 in the middle of the interval $[3, 7]$, the algorithm converges to the closest minimum point for function h and to a maximum point for the function g, i.e., it fails for this starting point. This gives rise to introducing the concept of a *region of attraction* of a minimum point x^*. A region of attraction of point x^* is the region of starting points x_0 where the local search procedure converges to point x^*. We elaborate this concept further in Section 4.2.4.

Table 4.2. Newton for functions g and h, $\alpha = 0.001$

		function g			function h		
k	x_k	$g'(x_k)$	$g''(x_k)$	$g(x_k)$	x_k	$h'(x_k)$	$h''(x_k)$	$h(x_k)$
0	5.000	-1.795	-4.934	1.301	5.000	-1.795	43.066	1.301
1	4.636	0.820	-7.815	1.511	5.042	0.018	43.953	1.264
2	4.741	-0.018	-8.012	1.553	5.041	0.000	43.944	1.264
3	4.739	0.000	-8.017	1.553	5.041	0.000	43.944	1.264

One can observe here when experimenting further, that when x_0 is close to a minimum (or maximum) point of g, the algorithm converges to that minimum (or maximum) point. Moreover, notice the effect of the safeguard to keep the iterates in interval $[3, 7]$. If $x_{k+1} < 3$, it is forced to a value of 3. One can find by experimentation that the left point $l = 3$ is also an attraction point of the algorithm for function g. Function h is piecewise convex, such that the algorithm always converges to the nearest minimum point.

4.2.3 Deterministic GO: Grid search, Piyavskii–Shubert

The aim of deterministic GO algorithms is to approach the optimum with a given certainty. We sketch two algorithms for the analysis of effectiveness and efficiency. The idea of reaching the optimum with an accuracy of ϵ can be done by so-called "everywhere dense sampling," as introduced in the literature on Global Optimization, e.g., Törn and Žilinskas (1989). In a rectangular domain this can be done by constructing a grid with a mesh of ϵ. By evaluating all

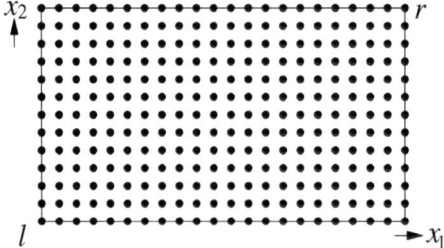

Fig. 4.4. Equidistant grid over rectangular feasible set

points on the grid, the best point found is a nice approximation of the global minimum point. The difficulty of GO is that even this best point found may be far away from the global minimum point, as the function may have a needle shape in another region in between the grid points. As shown in the literature, one can always construct a polynomial of sufficiently high degree, which fits all the evaluated points and has a minimum point more than ϵ away from the best point found. Actually, grid search is theoretically not effective if no further assumptions are posed on the optimization problem to be solved.

Let us have a look at the behavior of the algorithm for our two cases. For ease of formulation, we write down the grid algorithm for one-dimensional functions. The best function value found f^U is an upper bound for the minimum over the feasible set. We denote by x^U the corresponding best point found. The algorithm starts with the domain $[l, r]$ written as an interval and generates $M = \lceil (r - l)/\epsilon \rceil + 1$ grid points, where $\lceil x \rceil$ is the lowest integer greater than or equal to x.

Experimenting with test functions g and h gives reasonable results for $\epsilon = 0.01$, $(M = 401)$ and $\epsilon = 0.1$, $(M = 41)$. In both cases one finds an approximation x^U less than ϵ from the global minimum point. One knows

---

Algorithm 3 Grid($[l, r], f, \epsilon$)

$M = \lceil (r - l)/\epsilon \rceil + 1,\ f^U := \infty$
for ($k := 1$ to M) **do**
$\quad x_k := l + \frac{(k-1)(r-l)}{M-1}$
$\quad$ **if** ($f(x_k) < f^U$)
$\quad\quad f^U := f(x_k)$ and $x^U := x_k$
endfor

---

exactly how many function evaluations are required to reach this result in advance.

The efficiency of the algorithm in higher dimensions is also easy to establish. Given the lower left vector l and upper right vector r of a rectangular domain, one can easily determine how many grid coordinates $M_j, j = 1, \ldots, n$, in each direction should be taken and the total number of grid points is $\prod_j M_j$. This number is growing exponentially in the dimension n. As mentioned before, the effectiveness is not guaranteed in the sense of being closer than ϵ from a global minimum point, unless we make an assumption on the behavior of the function. A common assumption in the literature is Lipschitz continuity.

Definition 9. L is called a Lipschitz constant of f on X if

$$| f(x) - f(y) | \leq L \|x - y\|, \quad \forall x, y \in X.$$

In a practical sense it means that big jumps do not appear in the function value; slopes are bounded. With such an assumption, the δ-accuracy in the function space translates into an ϵ-accuracy in the x-space. Choosing $\epsilon = \delta/L$ gives that the best point x^U is in function value finally close to minimum point x^*:

$$| f^U - f^* | \leq L \|x^U - x^*\| \leq L\epsilon = \delta. \tag{4.8}$$

In higher dimension, one should be more exact in the choice of the distance norm $\|\cdot\|$. Here, for the one-dimensional examples we can focus on deriving the accuracy for our cases in a simple way. For a one-dimensional differentiable function f, L can be taken as

$$L = \max_{x \in X} | f'(x) |. \tag{4.9}$$

Using equation (4.9), one can now derive valid estimates for the example functions h and g. One can derive an over estimate L_g for the Lipschitz constant of g on $[3, 7]$ as

$$\max_{x \in [3,7]} | g'(x) | = \max_{x \in [3,7]} | \cos(x) + 3\cos(3x) + \tfrac{1}{x} |$$

$$\leq \max_{x \in [3,7]} \{ | \cos(x) | + | 3\cos(3x) | + | \tfrac{1}{x} | \} \tag{4.10}$$

$$\leq \max_{x \in [3,7]} | \cos(x) | + \max_{x \in [3,7]} | 3\cos(3x) | + \max_{x \in [3,7]} | \tfrac{1}{x} |$$

$$= 1 + 3 + \frac{1}{3} = L_g.$$

The estimate of L_h based on (4.7) is done by adding the maximum derivative of the bubble function $12 \times \frac{1}{2}$ to L_g for illustrative purposes rounded down to $L_h = 10$. We can now use (4.8) to derive a guarantee for the accuracy. One certainly arrives closer than $\delta = 0.01$ to the minimum in function value by taking a mesh size of $\epsilon = \frac{0.01}{4.33} = 0.0023$ for function g and taking $\epsilon = 0.001$ for function h. For the efficiency of grid search this means that reaching the δ-guarantee requires the evaluation of $M = 1733$ points for function g and $M = 4001$ points for function h. Note that due to the one-dimensional nature of the cases, ϵ can be taken twice as big, as the optimum point x^* is not further than half the mesh size from an evaluated point.

The main idea of most deterministic algorithms is not to generate and evaluate points everywhere dense, but to throw out those regions where the optimum cannot be situated. Given a Lipschitz constant, *Piyavskii and Shubert* independently constructed similar algorithms; see Shubert (1972) and Danilin and Piyavskii (1967). From the point of view of the graph of the function f to be minimized and an evaluated point (x_k, f_k), one can say that the region described by $\{(x, y)|y < f_k - L|x - x_k|\}$ cannot contain the optimum; the graph is above the function $f_k - L \mid x - x_k \mid$. Given a set of evaluated points $\{x_k\}$, one can construct a lower bounding function, a so-called sawtooth under-estimator that is given by $\varphi(x) = \max_k(f_k - L \mid x - x_k \mid)$ as illustrated by Figure 4.5. Given that we also have an upper bound f^U on the minimum of f being the best function value found thus far, one can say that the minimum point has to be in one of the shaded areas.

We will describe here the algorithm from a branch and bound point of view, where the subsets are defined by intervals $[l_p, r_p]$ and the endpoints are given by evaluated points. The index p is used to represent the intervals in

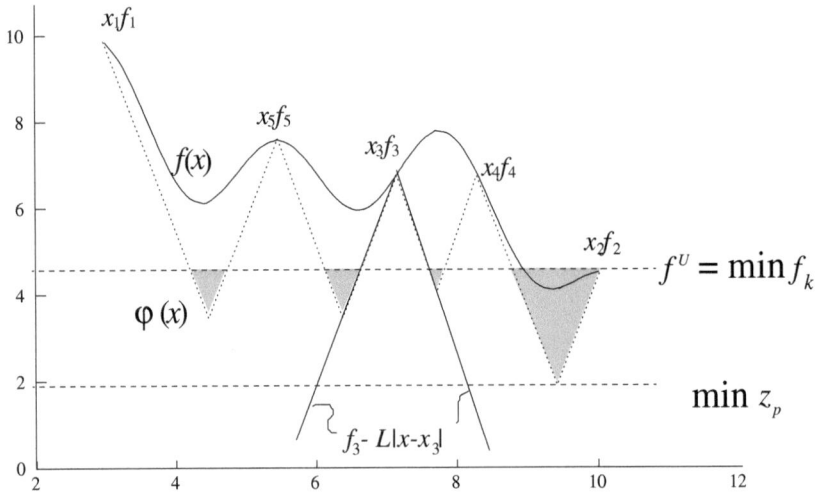

Fig. 4.5. Piyavskii–Shubert algorithm

Algorithm 4 PiyavShub($[l, r], f, L, \delta$)

Set $p := 1$, $l_1 := l$ and $r_1 := r$, $\Lambda := \{[l_1, r_1]\}$
$z_1 := \frac{f(l_1) + f(r_1)}{2} - \frac{L(r_1 - l_1)}{2}$, $f^U := \min\{f(l), f(r)\}$, $x^U := \text{argmin}\{f(l), f(r)\}$
while ($\Lambda \neq \emptyset$)
    remove an interval $[l_k, r_k]$ from Λ with $z_k = \min_p z_p$
    evaluate $f(m_k) := f(\frac{f(l_k) - f(r_k)}{2L} + \frac{r_k + l_k}{2})$
    if ($f(m_k) < f^U$)
        $f^U := f(m_k)$, $x^U := m_k$ and remove all C_p from Λ with $z_p > f^U - \delta$
    split $[l_k, r_k]$ into 2 new intervals $C_{p+1} := [l_k, m_k]$ and $C_{p+2} := [m_k, r_k]$
        with corresponding lower bounds z_{p+1} and z_{p+2}
    if ($z_{p+1} < f^U - \delta$), store C_{p+1} in Λ
    if ($z_{p+2} < f^U - \delta$), store C_{p+2} in Λ
    $p := p + 2$
endwhile

Λ. For each interval, a lower bound is given by

$$z_p = \frac{f(l_p) + f(r_p)}{2} - \frac{L(r_p - l_p)}{2}. \tag{4.11}$$

The gain with respect to grid search is that an interval can be thrown out as soon as $z_p > f^U$. Moreover, δ works as a stopping criterion as the algorithm implicitly (by not storing) compares the gap between f^U and $\min_p z_p$; stop if $(f^U - \min_p z_p) < \delta$. The algorithm proceeds by selecting the interval corresponding to $\min_p z_p$ (most promising) and splitting it over the minimum point of the sawtooth cover $\varphi(x)$ defined by

$$m_p = \frac{f(l_p) - f(r_p)}{2L} + \frac{r_p + l_p}{2} \tag{4.12}$$

being the next point to be evaluated. By continuing evaluating, splitting and throwing out intervals where the optimum cannot be, the stopping criterion is finally reached and we are certain to be closer than δ from f^* and therefore closer than $\epsilon = \delta/L$ from one of the global minimum points. The consequence of using such an algorithm, in contrast to the other algorithms, is that we now have to store information in a computer consisting of a list Λ of intervals. This computational effort is now added to that of evaluating sample points and doing intermediate calculations. This concept becomes more clear when running the algorithm on the test function g using an accuracy of $\delta = 0.01$. The Lipschitz constant $L_g = 4.2$ is used for illustrative purposes. As can be seen from Table 4.3, the algorithm is slowly converging. After some iterations, 15 intervals have been generated of which 6 are stored and 2 can be discarded due to the bounding; it has been proved that the minimum cannot be in the interval $[5.67, 7]$. The current estimate of the optimum is $x^U = 3.66$, $f^U = -0.2$ and the current lower bound is given by $\min_p z_p = -0.66$. Figure 4.6 illustrates the appearing binary structure of the search tree.

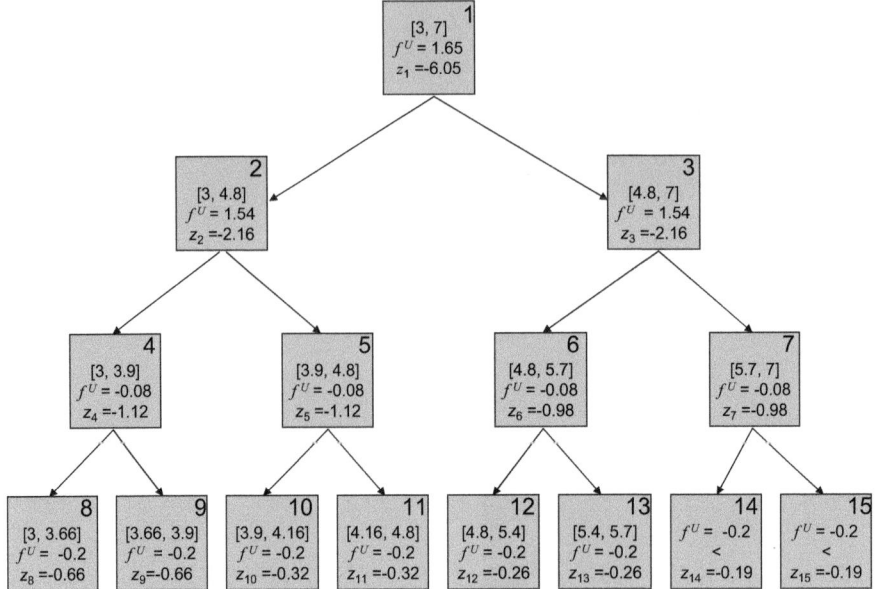

Fig. 4.6. Branch and bound tree of Piyavskii–Shubert for function g

The maximum computational effort with respect to storing intervals is reached when the branching proceeds and no parts can be thrown out; 2^K intervals appear at the bottom of the tree, where K is the depth of the tree. This mainly happens when the used Lipschitz parameter L drastically overestimates the maximum slope, or from another angle, the function is very flat compared to the used constant L. In that case, the function is evaluated in more than the M points of the regular grid. With a correct constant L, the number of evaluated points is less, as part of the domain can be discarded as illustrated here.

4.2.4 Stochastic GO: PRS, Multistart, Simulated Annealing

We consider stochastic methods as those algorithms that use (pseudo) random numbers to generate new trial points. For an overview of stochastic methods we refer to Boender and Romeijn (1995), Törn and Žilinskas (1989) and Törn et al. (1999). Two basic approaches in Global Optimization, Pure Random Search (PRS) and Multistart, are analyzed for the test cases. This is followed by a classical variant of Simulated Annealing, a so-called heuristic.

Pure Random Search (PRS) generates points uniformly over the domain and stores the point corresponding to the best value as the approximation of the global minimum point. The algorithm is popular as a reference algorithm as it can easily be analyzed. The question can now be how it behaves for our test cases g and h. The domain is clearly the interval $[3, 7]$, but

Table 4.3. Piyavskii–Shubert for function g, $\delta = 0.01$

p	l_p	r_p	$f(l_p)$	$f(r_p)$	m_p	z_p	f^U	x^U	
1	3.00	7.00	1.65	3.44	4.79	-5.85	1.65	3.00	split
2	3.00	4.79	1.65	1.54	3.91	-2.16	1.54	4.79	split
3	4.79	7.00	1.54	3.44	5.67	-2.16	1.54	4.79	split
4	3.00	3.91	1.65	-0.08	3.66	-1.12	-0.08	3.91	split
5	3.91	4.79	-0.08	1.54	4.15	-1.12	-0.08	3.91	split
6	4.79	5.67	1.54	0.20	5.39	-0.98	-0.08	3.91	split
7	5.67	7.00	0.20	3.44	5.95	-0.98	-0.08	3.91	split
8	3.00	3.66	1.65	-0.20	3.55	-0.66	-0.20	3.66	
9	3.66	3.91	-0.20	-0.08	3.77	-0.66	-0.20	3.66	
10	3.91	4.15	-0.08	0.47	3.96	-0.32	-0.20	3.66	
11	4.15	4.79	0.47	1.54	4.34	-0.32	-0.20	3.66	
12	4.79	5.39	1.54	0.46	5.22	-0.26	-0.20	3.66	
13	5.39	5.67	0.46	0.20	5.56	-0.26	-0.20	3.66	
14	5.67	5.95	0.20	0.61	5.76	-0.19	-0.20	3.66	discarded
15	5.95	7.00	0.61	3.44	6.14	-0.19	-0.20	3.66	discarded

Algorithm 5 PRS(X, f, N)

$f^U := \infty$
for $(k := 1$ to $N)$ **do**
    Generate x_k uniformly over X
    if $(f(x_k) < f^U)$
        $f^U := f(x_k)$ and $x^U := x_k$
endfor

---

what can be defined as the success region now? Let a success be defined as the case that one of the generated points is closer than $\epsilon = 0.01$ to the global minimum point. The probability we do NOT hit this region after $N = 50$ trials is $(3.98/4)^{50} \approx 0.78$. In the specific case, the size of the success region is namely 2ϵ and the size of the feasible area is 4. The probability of NOT hitting is $(1 - \frac{0.02}{4})$ and of NOT hitting 50 times is $(1 - \frac{0.02}{4})^{50}$. This means that the probability of success as efficiency indicator has a value of about 0.22 for both cases h and g.

A similar analysis can be done for determining the probability that the function value of PRS after $N = 50$ iterations is less than $f^* + \delta$ for $\delta = 0.01$. The usual tool in the analysis on the function space is to introduce $y = f(x)$ as a random variate representing the function value, where x is uniformly distributed over X. Value y has cumulative distribution function $\mu(y) = P(f(x) \leq y)$. We will elaborate on this in Chapter 7.

Keeping this in mind, analysis with so-called extreme-order statistics has shown that the outcome of PRS as record value of N points can be easily derived from $\mu(y)$. For a complete introduction to extreme-order statistics in optimization, we refer to Zhigljavsky (1991). Under mild assumptions it can be shown that $y_{(1)} = \min\{f(x_1), \ldots, f(x_N)\}$ has the distribution function

$F_{(1)}(y) = 1 - (1 - \mu(y))^N$. This means that for the question about the probability that $y_{(1)} \leq f^* + \delta$, we do not have to know the complete distribution function μ, but only the probability mass $\mu(f^* + \delta)$ of the success level set where $f(x) \leq f^* + \delta$, i.e., the probability that one sample point hits this low level set. Here the two test cases differ considerably. One can verify that the level set of the more smooth test function g is about 0.09 wide, whereas that of function h is only 0.04 wide for a δ of 0.01. This means that the probability of PRS to reach a level below $f^* + \delta$ after 50 evaluations for function g is $1 - (1 - \frac{0.09}{4})^{50} = 0.68$, whereas the same probability for function h is $1 - (1 - \frac{0.04}{4})^{50} = 0.40$.

An early observation based on the extreme-order statistic analysis is due to Karnopp (1963). Surprisingly enough, Karnopp showed that the probability of finding a better function value with one draw more after N points have been generated, is $\frac{1}{N+1}$, independent of the problem to be solved. Generating K more points increases the probability to $\frac{K}{N+K}$. The derivation which also can be found in Törn and Žilinskas (1989) is based on extreme-order statistics and only requires μ not to behave too strange, e.g., μ is continuous such that f should not have plateaus on the domain X.

Stochastic algorithms show something which in the literature is called the *infinite effort property*. This means that if one proceeds long enough (read $N \to \infty$), in the end the global optimum is found. The problem with such a concept is that infinity can be pretty far away. Moreover, we have seen in the earlier analyses that the probability of reaching what one wants, depends considerably on the size of the success region. One classical way of increasing the probability of reaching an optimum is to use (nonlinear optimization) local searches. This method is called *Multistart*.

Define a local optimization routine $LS(x) : X \to X$ as a procedure which given a starting point returns a point in the domain that approximates a local minimum point. As an example, one can consider the Newton method of Section 4.2.2. Multistart generates convergence points of a local optimization routine from randomly generated starting points.

Algorithm 6 Multistart(X, f, LS, N)

$f^U := \infty$
for $(k := 1$ to $N)$ **do**
    Generate x uniformly over X
    $x_k := LS(x)$
    if $(f(x_k) < f^U)$
        $f^U := f(x_k)$ and $x^U := x_k$
endfor

Note that the number of iterations N is not comparable with that in PRS, as every local search requires several function evaluations. Let us for the example cases assume that the Newton algorithm requires 5 function evaluations to detect an attraction point, as is also implied by Table 4.2. As we were using $N = 50$ function evaluations to assess the success of PRS on the test cases, we will use $N = 10$ iterations for Multistart. In order to determine a similar probability of success, one should find the relative size of the region of attraction of the global minimum point. Note again that the Newton algorithm does not always converge to the nearest optimum; it only converges to a minimum point in a convex region around it.

For function g, the region of attraction of the global minimum is not easy to determine. It consists of a range of about 0.8 on the feasible area of size 4, such that the probability of one random starting point leading to success is $0.8/4 = 0.2$. For function h, the good region of attraction is simply the bubble of size 0.25 around the global minimum point, such that the probability of finding the global minimum in one iteration is about 0.06. Reaching the optimum after $N = 10$ restarts is $1 - 0.8^{10} \approx 0.89$ for g and $1 - 0.94^{10} \approx 0.48$ for h. In both examples, the probability of success is larger than that of PRS.

As sketched so far, the algorithms of Pure Random Search and Multistart have been analyzed widely in the literature of GO. Algorithms that are far less easy to analyze, but very popular in applications, are the collection of so-called meta-heuristics. This term was introduced by Fred Glover in Glover (1986) and includes simulated annealing, evolutionary algorithms, genetic algorithms, tabu search, and all the fantasy names derived from crossovers of the other names. Originally these algorithms were not only aimed at continuous optimization problems; see Aarts and Lenstra (1997). An interesting research question is whether they are really better than combining classical ideas of random search and nonlinear optimization local searches. We discuss here a variant of simulated annealing , a concept that also got attention in the GO literature; see Romeijn (1992). Simulated annealing describes a sampling process in the decision space where new sample points are generated from a so-called neighborhood of the current iterate. The new sample point is always accepted when it is better and with a certain probability when it is worse. The probability depends on the so-called temperature that is decreasing (cooling) during the iterations.

The algorithm contains the parameter CR representing the *cooling rate* with which the temperature variable decreases. A fixed value of 1000 was taken for the initial temperature to avoid creating another algorithm parameter. The algorithm accepts a worse point depending on how much it is worse and the development of the algorithm. This is a generic concept in simulated annealing. There are several ways to implement the concept of "sample from neighborhood." In one dimension one would perceive intuitively a neighborhood of x_k in real space $[x_k - \epsilon, x_k + \epsilon]$, which can be found in many algorithms; e.g., see Baritompa et al. (2005). As such heuristics were originally not aimed at continuous optimization problems, but at integer problems, one of the first approaches was the coding of continuous variables in bitstrings. For the illustrations, we elaborate this idea for the test case. Each point $x \in [3, 7]$ is

---

Algorithm 7 SA(X, f, CR, N)

$f^U := \infty$, $T_1 := 1000$
Generate x_1 uniformly over X
for ($k := 1$ to N) **do**
    Generate x from a neighborhood of x_k
    if ($f(x) < f(x_k)$)
        $x_{k+1} := x$
        if ($f(x) < f^U$)
            $f^U := f(x)$ and $x^U := x$
    else with probability $e^{\frac{f(x_k)-f(x)}{T_k}}$ let $x_{k+1} := x$
    $T_{k+1} := CR \times T_k$
endfor

---

represented by a bitstring $(B_1, \ldots, B_9) \in \{0,1\}^9$, where

$$x = 3 + 4\frac{\sum_{i=1}^9 B_i 2^{i-1}}{511}. \qquad (4.13)$$

Formula (4.13) describes a regular grid over the interval, where each of the $M = 512$ bitstrings is one of the grid points, such that the mesh size is $\frac{4}{511}$. The sampling from a neighborhood of a point x is done by flipping at random one of its bit variables B_i from a value of 0 to 1, or the other way around. Notice that by doing so, the generated point is not necessarily in what one would perceive as a neighborhood in continuous space. The question is therefore, whether the described SA variant will perform better than an algorithm where the new sample point does not depend on the current iterate, PRS. To test this, a figure is introduced that is quite common in experimenting with meta-heuristics. It is a graph with the effort on the x-axis and the reached success on the y-axis. The GO literature often looks at the two criteria effectiveness and efficiency separately. Figure 4.7 being a special case of what in Baritompa and Hendrix (2005) was called the *performance graph*, gives a trade-off between the two main criteria. One can also consider the x-axis to give a budget with which one has to reach a level as low as possible; see Hendrix and Roosma (1996). In this way one can change the search strategy depending on the amount of available running time. The figure suggests, for instance, that a high cooling rate CR (the process looks like PRS) does better for a lower number of function values and worse for a higher number of function values.

Figure 4.7 gives an estimation of the expected level one can reach by running SA on function g. Implicitly it says the user wants to reach a low function value; not necessarily a global minimum point. Theoretically, one can derive the expected level analytically by considering the process from a Markov chain perspective; see, e.g., Bulger and Wood (1998). However, usually the estimation is done empirically and the figure is therefore very common in metaheuristic approaches. The reader will not be surprised that

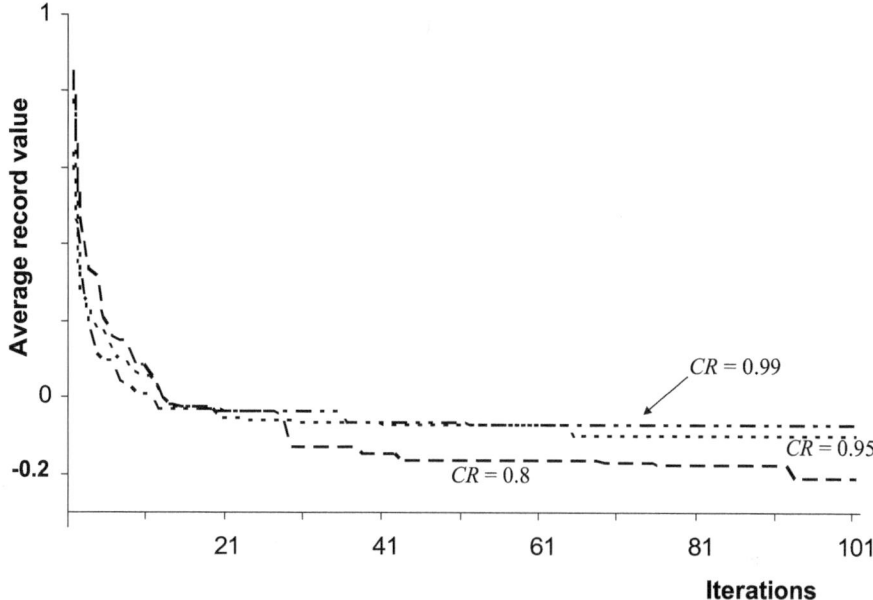

Fig. 4.7. Average record value over 10 runs of SA, three different values of CR for a given amount of function evaluations, test case g

the figure looks similar for function h, as the number of local optima is not relevant for the bitstring perspective and the function value distribution is similar. Theoretically, one can also derive the expected value of the minimum function value reached by PRS. It is easier to consider the theoretical behavior from the perspective where success is defined Boolean, as has been done so far.

Let us consider again the situation that the algorithm reaches the global optimum as success. For stochastic algorithms we are interested in the probability of success. Define reaching the optimum again as finding a point with function value closer than $\delta = 0.01$ to the minimum. For function g this is about 2.2% (11 out of the 512 bitstrings) of the domain. For PRS, one can determine the probability of success as $P_{PRS}(N) = 1 - 0.978^N$. For SA this is much harder to determine, but one can estimate the probability of success empirically. The result is the performance graph in Figure 4.8. Let us have a look at the figure critically. In fact it suggests that PRS is doing as well as the SA algorithms. As this is verifying a hypothesis (not falsifying), this is a reason to be suspicious. The following critical remarks can be made.

- The 10 runs are enough to illustrate how the performance can be estimated, but are too low to discriminate between methods. Perhaps the author has even selected a set of runs which fits the hypothesis nicely.
- One can choose the scale of the axes to focus on an effect. In this case, one can observe that up to 40 iterations, PRS does not look better than

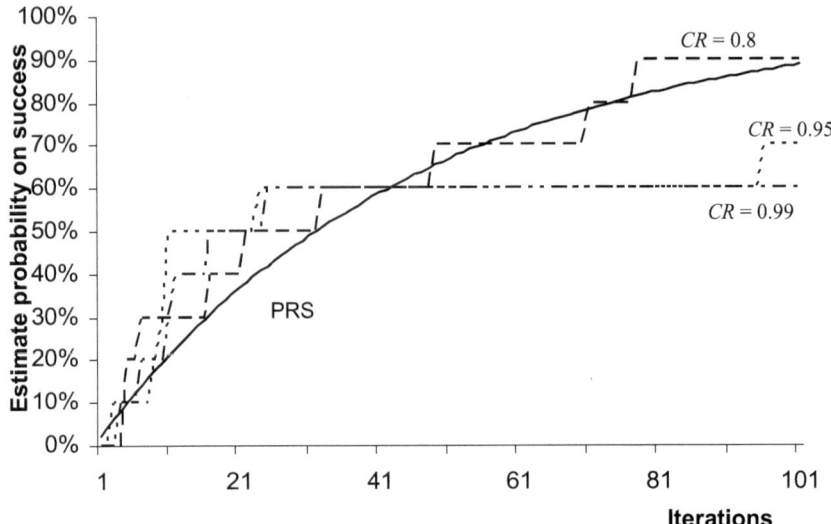

Fig. 4.8. Estimate of probability of reaching the minimum for PRS and three cooling rate (CR) values for SA (average over 10 runs) on function g given a number of function evaluations

the SA variants. By choosing the x-axis to run to 100 iterations, it looks much better.

- The graph has been depicted for function g, but not for function h, where the size of the success region is twice as small. One can verify that, in the given range, the SA variants nearly always do better.

. This brings us to the general scientific remark, that all results should be described in such a way that they can be *reproduced*, i.e., one should be able to repeat the experiment. For the exercises reported in this section, this is relatively simple. Spreadsheet calculations and for instance MATLAB implementations can easily be made.

4.3 Investigating algorithms

In Section 4.2, we have seen how several algorithms behave on two test cases. What did we learn from that? How do we carry out the investigation systematically? Figure 4.9 depicts some relevant aspects. All aspects should be considered together. The following steps are described in Baritompa and Hendrix (2005).

1. Formulation of performance criteria.
2. Description of the algorithm(s) under investigation.
3. Selection of appropriate algorithm parameters.

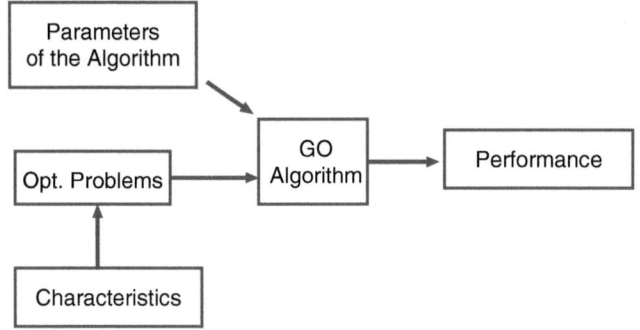

Fig. 4.9. Aspects of investigating Global Optimization algorithms

4. Production of test functions (instances, special cases) corresponding to certain landscape *structures or characteristics*.
5. Analysis of its theoretical performance, or empirical testing.

Several criteria and performance indicators have been sketched: to get a low value, to reach a local minimum point, a high probability to hit an ϵ neighborhood of a global minimum point, obtain a guarantee to be less than δ from the minimum function value, etc. Several classes of algorithms have been outlined. The number of parameters has been kept low, a Lipschitz constant L, the number of iterations N, cooling rate CR, stopping accuracy α. Many modern heuristics contain so many tuning parameters that it is hard to determine the effect of their value on the performance of the algorithm.

Only two test functions were introduced to experiment with. The main difference between them is the number of optima, which in the literature is seen as an important characteristic. However, in the illustrations, the number of optima was only important for the performance of Multistart. The piecewise convexity appeared important for the local search Newton algorithm and further the difference in size of the δ-level set was of especial importance for the probability of success of stochastic algorithms. It teaches us that, in a research setting, one should think carefully, make hypotheses and design corresponding experiments, to determine which characteristics of test functions are relevant for the algorithm under investigation.

4.3.1 Characteristics

In Baritompa et al. (2005), an attempt is made to analyze the interrelation between the characteristics, called "landscape" in their view, of the test cases and the behavior of the algorithms. What we see experimentally is that often an algorithm is run over several test functions and its performance compared to other algorithms and/or other parameter settings. To understand behavior, we need to study the relationships to characteristics (landscapes) of test functions. The main question is how to define appropriate characteristics. We will discuss some ideas which appear in the literature. The main idea, as illustrated before, is that relevant characteristics depend on the type of algorithm as well as on the performance measure. Extending previous stochastic examples to more dimensions, it is only the relative size of the sought-for success region that matters, characteristics such as the number of optima, the shape of regions of attraction, the form of the level sets, barriers in the landscape do not matter.

It is also important to vary the test cases systematically between the extreme cases, in order to understand how algorithms behave. In an experimental setting, depending on what one measures, one tries to design experiments which yield as much information as possible. To derive analytical results, it is not uncommon to make highly specific assumptions which make the analysis tractable. In the GO literature, the following classes of problems with respect to available information are distinguished.

- Black-box (or oracle) case: in this case it is assumed that nothing is known about the function to be optimized. Often the feasible set is defined as a box, but information about the objective function can only be obtained by evaluating the function at feasible points.
- Gray-box case: something is known about the function, but the explicit form is not necessarily given. We may have a lower bound on the function value or on the number of global and/or local optima. As has proved useful for deterministic methods, we may have structural information such as: a concave function, a known Lipschitz constant. Stochastic methods often do not require this type of information, but this information may be used to derive analytical or experimental results.
- White-box case: explicit analytical expressions of the problem to be solved are assumed to be available. Specifically so-called interval arithmetic algorithms require this point of view on the problem to be solved.

When looking at the structure of the instances for which one studies the behavior of the algorithm, we should keep two things in mind.

- In experiments, the researcher can try to influence the characteristics of the test cases such that the effect on what is measured is as big as possible. Note that the experimentalist knows the structure in advance, but the algorithm does not.
- The algorithm can try to generate information which tells it about the structure of the problems. We will enumerate some information which can be measured in the black-box case.

Considering the lists of test functions initially in the literature (e.g., Törn and Žilinskas, 1989) and later on the Internet, one can see as characteristics the number of global minimum points, the number of local minimum points and the dimension of the problem.

A difficulty in the analysis of a GO algorithm in the multiextremal case is that everything seems to influence behavior: The orientation of components of lower level sets with respect to each other determines how iterates can jump from one place to the other. The number of local optima up in the "hills" determines how algorithms may get stuck in local optima. The difference between the global minimum and the next lowest minimum affects the ability to detect the global minimum point. The steepness around minimum points, valleys, creeks, etc. which determine the landscape influences the success. However, we stress, as shown by the examples, the problem *characteristics* which are important for the behavior depend on the *type of algorithm* and the *performance criteria* that measure the success.

In general, stochastic algorithms require no structural information about the problem. However, one can adapt algorithms to make use of structure information. Moreover, one should notice that even if structural information is not available, other so-called *value information* becomes available when running algorithms: the number of local optima found thus far, the average number of function evaluations necessary for one local search, best function value found, and the behavior of the local search, etc. Such indicators can be measured empirically and can be used to get insight into what factors determine the behavior of a particular algorithm and perhaps can be used to improve the performance of an algorithm. From the perspective of designing algorithms, running them empirically may *generate information* about the landscape of the problem to be solved. The following list of information can be collected during running of a stochastic GO algorithm on a black-box case; see Hendrix (1998):

- Graphical information on the decision space.
- Current function value.
- Best function value found so far (record).
- Number of evaluations in the current local phase.
- Number of optima found.
- Number of times each detected minimum point is found.
- Estimates of the time of one function evaluation.
- Estimates of the number of function evaluations for one local search.
- Indicators on the likelihood to have found an optimum solution.

For the likelihood indicator, a probability model is needed. Measuring and using the information in the algorithm, usually leads to more extended algorithms, called "adaptive." Often, these have additional parameters complicating the analysis of what are good parameter settings.

4.3.2 Comparison of algorithms

When comparing algorithms, a specific algorithm is *dominated* if there is another algorithm which performs better (e.g., has a higher probability performance graph) in all possible cases under consideration. Usually, however, one algorithm runs better on some cases and another on other cases.

So basically, the performance of algorithms can be compared on the same test function, or preferably for many test functions with the same characteristic, measured by the only parameter that matters for the performance of the compared algorithms. As we have seen, it may be very hard to discover such characteristics. The following principles can be useful:

- Comparability: Compared algorithms should make use of the same type of (structural) information (same stopping criteria, accuracies, etc.).
- Simple references: It is wise to include in the comparison simple benchmark algorithms such as Pure Random Search, Multistart and Grid Search in order not to let analysis of the outcomes get lost in parameter tuning and complicated schemes.
- Reproducibility: In principle, the description of the method that is used to generate the results has to be so complete that someone else can repeat the exercise obtaining similar results (not necessarily the same).

Often in applied literature, we see algorithms used for solving "practical" problems; see, e.g., Ali et al. (1997), Hendrix (1998), and Pintér (1996) for extensive studies. As such this illustrates the relevance of the study on Global Optimization algorithms. In papers where one practical problem is approached with one (type of) algorithm, the reference is lacking. First of all we should know the performance of simple benchmark algorithms on that problem. Second, if structure information is lacking, we do not learn a lot about the performance of the algorithm under study on a class of optimization problems. We should keep in mind this is only one problem and up to now, it does not represent all possible practical problems.

4.4 Summary and discussion points

- Results of investigation of GO algorithms consist of a description of the performance of algorithms (parameter settings) depending on characteristics of test functions or function classes.
- To obtain good performance criteria, one needs to identify the target of an assumed user and to define what is considered a success.
- The performance graph is a useful instrument for comparing performance.
- The relevant characteristics of the test cases depend on the type of algorithm and performance criterion under consideration.
- Algorithms to be comparable must make use of same information, accuracies, principles, etc.
- It is wise to compare performance to simple benchmark algorithms like Grid Search, PRS and Multistart in a study on a GO algorithm.
- Description of the research path followed should allow reproduction of experiments.

4.5 Exercises

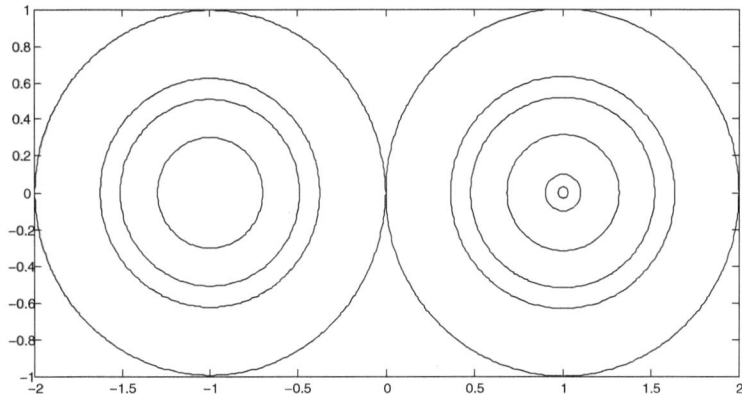

Fig. 4.10. Contour lines of bi-spherical problem

1. Given function $f(x) = \frac{1}{2}x + \frac{1}{x}$ on interval $X = [1, 4]$.
   (a) Perform three iterations of the bisection algorithm.
   (b) How many iterations are required to reach an accuracy of $\epsilon = 10^{-4}$?

2. With Pure Random Search we want to generate a point on interval $X = [1, 4]$ for which $f(x) = x^2 - 4x + 6$ has a low function value.
   (a) Determine the level set $S_\delta = \{x \in X | f(x) \le f^* + \delta\}$ for $\delta = 0.01$.
   (b) Determine the probability a random generated point on X is in S_δ.
   (c) Determine the probability a point in S_δ is found after 10 iterations.

3. Given $f(x) = \min\{(x_1 - 1)^2, (x_1 + 1)^2 + 0.1\} + x_2^2$, a so-called bi-spherical objective function over feasible set defined by the box constraints $-2 \le x_1 \le 2$ and $-1 \le x_2 \le 1$, i.e., $X = [-2, 2] \times [-1, 1]$. See Figure 4.10.
   (a) Estimate the relative volume of level set $S_{0.01} = \{x \in X | f(x) \le 0.01\}$.
   (b) How would you generate grid points over X, such that the lowest point is expected to be closer than $\epsilon = 0.01$ to the global minimum point? How many points do you have to evaluate accordingly?
   (c) Determine the probability that Pure Random Search has found a point in $S_{0.01}$ after $N = 100$ trials.

4. Consider minimization of $f(x) = \sin(x) + \sin(3x) + \ln(x), x \in [3, 7]$ via Algorithm 8 with a population size $N = 10$.
   (a) How many points have been evaluated after $k = 31$ iterations?
   (b) Execute Algorithm 8 for $f(x)$ over $[3, 7]$ taking $\epsilon = 0.1$.
   (c) Is it possible that the algorithm does not converge for this case?
   (d) How to estimate the number of used function evaluations numerically? How many evaluations does the algorithm require on average up to convergence?

Algorithm 8 Stochastic Population(X, f, N, ϵ)

Generate a set P of N points uniformly over feasible set X
Evaluate all points in P; $k := 0$
Determine $f^U := f(x^U) = \max_{p \in P} f(p)$ and $f^u := \min_{p \in P} f(p)$
while $(f^U - f^u > \epsilon)$
  $k := k + 1$
  Select at random two parents $p_1, p_2 \in P$
  Determine midpoint $c := \frac{1}{2}(p_1 + p_2)$
  Take x_k uniformly (chance of $\frac{1}{3}$) from $\{c, \frac{3}{2}p_1 - \frac{1}{2}c, \frac{3}{2}p_2 - \frac{1}{2}c\}$
  Evaluate $f(x_k)$;
  if $(f(x_k) < f^U)$, replace x^U by x_k in P
  Determine $f^U := f(x^U) = \max_{p \in P} f(p)$ and $f^u := \min_{p \in P} f(p)$
endwhile

(e) How can the algorithm be modified in such a way that the population stays in the feasible set?

5

Nonlinear Programming algorithms

5.1 Introduction

This chapter describes algorithms that have been specifically designed for finding optima of Nonlinear Programming (NLP) problems.

5.1.1 General NLP problem

The generic NLP problem has been introduced in Chapter 1:

$$\begin{aligned} &\min\ f(x) \\ &\text{s.t.}\ \ g_i(x) \le 0 \text{ for some properties } i, \text{ inequality constraints,} \\ &\qquad g_i(x) = 0 \text{ for some properties } i, \text{ equality constraints.} \end{aligned} \qquad (5.1)$$

where x is moving in a continuous way in the feasible set X that is defined by the inequality and equality constraints. An important distinction from the perspective of the algorithms is whether derivative information is available on the functions f and g_i. We talk about first-order derivative information if the vector of partial derivatives, called the gradient, is available in each feasible point.

The most important distinction is that between *smooth* and *nonsmooth* optimization. If the functions f and g are continuously differentiable, one speaks of smooth optimization. In many practical models, the functions are not everywhere differentiable as illustrated in Chapter 2, e.g., Figure 2.3.

5.1.2 Algorithms

In general one would try to find "a" or "the" optimum with the aid of software called a solver, which is an implementation of an algorithm. For solvers related to modeling software, see, e.g., the GAMS-software (www.gams.com), AMPL (www.ampl.com), Lingo (www.lindo.com) and AIMMS (www.aimms.com).

E.M.T. Hendrix and B.G.-Tóth, *Introduction to Nonlinear and Global Optimization,* 91
Springer Optimization and Its Applications 37, DOI 10.1007/978-0-387-88670-1_5,
© Springer Science+Business Media, LLC 2010

Following the generic description of Törn and Žilinskas (1989), an NLP algorithm can be described as

$$x_{k+1} = Alg(x_k, x_{k-1}, \ldots, x_0) \tag{5.2}$$

where index k is the iteration counter. Formula (5.2) represents the idea that a next point x_{k+1} is generated based on the information in all former points $x_k, x_{k-1}, \ldots, x_0$, where x_0 is called the starting point. The aim of an NLP algorithm is to detect a (local) optimum point x^* given the starting point x_0. Usually one is satisfied if convergence takes place in the sense of $x_k \to x^*$ and/or $f(x_k) \to f^*$. Besides the classification of using derivative information or not, another distinction is whether an algorithm aims for constrained optimization or unconstrained optimization. We talk about constrained optimization if at least one of the constraints is expected to be binding in the optimum, i.e., $g_i(x^*) = 0$ for at least one constraint i. Otherwise, the constraints are either absent or can be ignored. We call this unconstrained optimization.

In the literature on NLP algorithms (e.g., Scales, 1985; Gill et al., 1981), the basic cycle of Algorithm 9 is used in nearly each unconstrained NLP algorithm.

Algorithm 9 GeneralNLP(f, x_0)

Set $k := 0$
while passing stopping criterion
    $k := k + 1$
    determine search direction r_k
    determine step size λ_k along line $x_k + \lambda r_k$
    next iterate is $x_{k+1} := x_k + \lambda_k r_k$
endwhile

The determination of the step size λ_k is done in many algorithms by running an algorithm for minimizing the one-dimensional function $\varphi_{r_k}(\lambda) = f(x_k + \lambda r_k)$. This is called line minimization or line search, i.e., f is minimized over the line $x_k + \lambda r_k$. In the discussion of algorithms, we first focus on minimizing functions in one variable in Section 5.2. They can be used for line minimization. In Section 5.3, algorithms are discussed that require no derivative information. We will also introduce a popular algorithm that does not follow the scheme of Algorithm 9. Algorithms that require derivative information can be found in Section 5.4. A large class of problems is due to nonlinear regression problems. Specific algorithms for this class are outlined in Section 5.5. Finally, Section 5.6 outlines several concepts that are used to solve NLP problems with constraints.

5.2 Minimizing functions of one variable

Two concepts are important in finding a minimum of $f : \mathbb{R} \rightarrow \mathbb{R}$; that of interval reduction and that of interpolation. Interval reduction enhances determining an initial interval and shrinking it iteratively such that it includes a minimum point. Interpolation makes use of information of function value and/or higher-order derivatives. The principle is to fit an approximating function and to use its minimum point as a next iterate. Practical algorithms usually combine these two concepts. Several basic algorithms are described.

5.2.1 Bracketing

In order to determine an interval that contains an internal optimum given starting point x_0, bracketing is used. It iteratively walks further until we are certain to have an interval (bracket) $[a, b]$ that includes an interior minimum point. The algorithm enlarges the initial interval with endpoints x_0 and $x_0 \pm$

Algorithm 10 Bracket(f, x_0, ϵ, a, b)

Set $k := 1$, $\varrho = \frac{2}{\sqrt{5}-1}$
if $(f(x_0 + \epsilon) < f(x_0))$
    $x_1 := x_0 + \epsilon$
else if $(f(x_0 - \epsilon) < f(x_0))$
    $x_1 := x_0 - \epsilon$
else STOP; x_0 is optimal
repeat
    $k := k + 1$
    $x_k := x_{k-1} + \varrho(x_{k-1} - x_{k-2})$
until $(f(x_k) > f(x_{k-1}))$
$a := \min\{x_k, x_{k-2}\}$
$b := \max\{x_k, x_{k-2}\}$

ϵ with a step that becomes each iteration a factor $\varrho > 1$ bigger. Later, in Section 5.2.3 it will be explained why exactly the choice $\varrho = \frac{2}{\sqrt{5}-1}$ is convenient. It stops when finally x_{k-1} has a lower function value than x_k as well as x_{k-2}.

Example 5.1. The bracketing algorithm is run on the function $f(x) = x + \frac{16}{x+1}$ with starting point $x_0 = 0$ and accuracy $\epsilon = 0.1$. The initial interval $[0, 0.1]$ is iteratively enlarged represented by $[x_{k-2}, x_k]$ and walks in the decreasing direction. After seven iterations, the interval $[1.633, 4.536]$ certainly contains a minimum point as there exists an interior point $x_{k-1} = 2.742$ with a function value lower than the endpoints of the interval; $f(2.742) < f(1.633)$ and $f(2.742) < f(4.536)$.

Table 5.1. Bracketing for $f(x) = x + \frac{16}{x+1}$, $x_0 = 0$, $\epsilon = 0.1$

k	x_{k-2}	x_k	$f(x_k)$
0		0.000	16.00
1		0.100	14.65
2	0.000	0.261	12.94
3	0.100	0.524	11.03
4	0.262	0.947	9.16
5	0.524	1.633	7.71
6	0.947	2.742	7.02
7	1.633	4.536	7.43

The idea of interval reduction techniques is now to reduce an initial interval that is known to contain a minimum point and to shrink it to a tiny interval enclosing the minimum point. One such method is bisection.

5.2.2 Bisection

The algorithm departs from a starting interval $[a, b]$ that is halved iteratively based on the sign of the derivative in the midpoint. This means that the method is in principle only applicable when the derivative is available at the generated midpoints. The point x_k converges to a minimum point within the interval $[a, b]$. If the interval contains only one minimum point, it converges to that. At each step, the size of the interval is halved and in the end, we are

---

Algorithm 11 Bisect$([a, b], f, \epsilon)$

Set $k := 0$, $a_0 := a$ and $b_0 := b$
while $(b_k - a_k > \epsilon)$
$\quad x_k := \frac{a_k + b_k}{2}$
$\quad$ **if** $f'(x_k) < 0$
$\quad\quad a_{k+1} := x_k$ and $b_{k+1} := b_k$
$\quad$ **else**
$\quad\quad a_{k+1} := a_k$ and $b_{k+1} := x_k$
$\quad k := k + 1$
endwhile

---

certain that the current iterate x_k is not further away than ϵ from a minimum point. It is relatively easy to determine for this algorithm how many iterations corresponding to (derivative) function evaluations are necessary to come closer than ϵ to a minimum point. Since $| b_{k+1} - a_{k+1} | = \frac{1}{2} | b_k - a_k |$, the number of iterations necessary for reaching ϵ-convergence is

$$| b_k - a_k | = (\tfrac{1}{2})^k | b_0 - a_0 | < \epsilon \quad \Rightarrow$$
$$(\tfrac{1}{2})^k < \tfrac{\epsilon}{|b_0 - a_0|} \quad \Rightarrow \quad k > \frac{\ln \epsilon - \ln |b_0 - a_0|}{\ln \frac{1}{2}}.$$

Table 5.2. Bisection for $f(x) = x + \frac{16}{x+1}$, $[a_0, b_0] = [2, 4.5]$, $\epsilon = 0.01$

k	a_k	b_k	x_k	$f(x_k)$	$f'(x_k)$
0	2.000	4.500	3.250	7.0147	0.114
1	2.000	3.250	2.625	7.0388	-0.218
2	2.625	3.250	2.938	7.0010	-0.032
3	2.938	3.250	3.094	7.0021	0.045
4	2.938	3.094	3.016	7.0001	0.008
5	2.938	3.016	2.977	7.0001	-0.012
6	2.977	3.016	2.996	7.0000	-0.002
7	2.996	3.016	3.006	7.0000	0.003
8	2.996	3.006	3.001	7.0000	0.000

For instance, $b_k - a_k = 4$ requires at least nine iterations to reach an accuracy of $\epsilon = 0.01$.

Example 5.2. The bisection algorithm is run on the function $f(x) = x + \frac{16}{x+1}$ with starting interval $[2, 4.5]$ and accuracy $\epsilon = 0.01$. The interval $[a_k, b_k]$ is slowly closing around the minimum point $x^* = 3$ which is approached by x_k. One can observe that $f(x_k)$ is converging fast to $f(x^*) = 7$. A stopping criterion on convergence of the function value, $\mid f(x_k) - f(x_{k-1}) \mid$, would probably have stopped the algorithm earlier. The example also shows that the focus of the algorithm is on approximating a point x^* where the derivative is zero, $f'(x^*) = 0$.

The algorithm typically uses derivative information. Usually the efficiency of an algorithm is measured by the number of function evaluations necessary to reach the goal of the algorithm. If the derivative is not analytically or computationally available, one has to evaluate in each iteration two points, x_k and $x_k + \delta$, where δ is a small accuracy number such as 0.0001. Evaluating in each iteration two points, leads to a reduction of the interval to its half at each iteration.

Interval reduction methods usually use the function value of two interior points in the interval to decide the direction in which to reduce it. One elegant way is to recycle one of the evaluated points and to use it in the next iterations. This can be done by using the so-called *Golden Section* rule.

5.2.3 Golden Section search

This method uses two evaluated points l (left) and r (right) in the interval $[a_k, b_k]$, that are located in such a way that one of the points can be used again in the next iteration. The idea is sketched in Figure 5.1. The evaluation points l and r are located with fraction τ in such a way that $l = a + (1 - \tau)(b - a)$ and $r = a + \tau(b - a)$. Equating in Figure 5.1 the next right point to the old left point gives the equation $\tau^2 = 1 - \tau$. The solution is the so-called Golden Section number $\tau = \frac{\sqrt{5}-1}{2} \approx 0.618$.

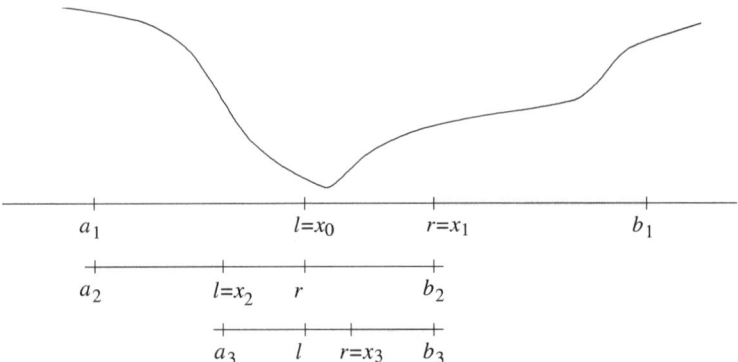

Fig. 5.1. Golden Section search

This value also corresponds to the value ϱ used in the Bracketing algorithm in the following way. Using the outcomes of the Bracketing algorithm as input into the Golden Section search as $[a, b]$ gives that the point x_{k-1} (of algorithm Bracket) corresponds to x_0 (in algorithm Goldsect). This means that it does not have to be evaluated again.

Example 5.3. The Golden Section search is run on the function $f(x) = x + \frac{16}{x+1}$ with starting interval $[2, 4.5]$ and accuracy $\epsilon = 0.1$ The interval $[a_k, b_k]$ encloses the minimum point $x^* = 3$. Notice that the interval is shrinking slower than by bisection, as $\mid b_{k+1} - a_{k+1} \mid = \tau \mid b_k - a_k \mid = \tau^{k-1} \mid b_1 - a_1 \mid$. After eight iterations the reached accuracy is less than by bisection, although for this case x_k approaches the minimum very well. On the other hand, only one function evaluation is required at each iteration.

Algorithm 12 Goldsect($[a, b], f, \epsilon$)

Set $k := 1$, $a_1 := a$ and $b_1 := b$, $\tau := \frac{\sqrt{5}-1}{2}$
$l := x_0 := a + (1 - \tau)(b - a)$, $r = x_1 := a + \tau(b - a)$
Evaluate $f(l) := f(x_0)$
repeat
    Evaluate $f(x_k)$
    if $(f(r) < f(l))$
        $a_{k+1} := l$, $b_{k+1} := b_k$, $l := r$
        $r := x_{k+1} := a_{k+1} + \tau(b_{k+1} - a_{k+1})$
    else
        $a_{k+1} := a_k$, $b_{k+1} := r$, $r := l$
        $l := x_{k+1} := a_{k+1} + (1 - \tau)(b_{k+1} - a_{k+1})$
    $k := k + 1$
until $(b_k - a_k < \epsilon)$

Table 5.3. Golden Section search for $f(x) = x + \frac{16}{x+1}$, $[a_0, b_0] = [2, 4.5]$, $\epsilon = 0.1$

k	a_k	b_k	x_k	$f(x_k)$
0			2.955	7.0005
1	2.000	4.500	3.545	7.0654
2	2.000	3.545	2.590	7.0468
3	2.590	3.545	3.180	7.0078
4	2.590	3.180	2.816	7.0089
5	2.816	3.180	3.041	7.0004
6	2.955	3.180	3.094	7.0022
7	2.955	3.094	3.008	7.0000
8	2.955	3.041	2.988	7.0000

5.2.4 Quadratic interpolation

The interval reduction techniques discussed so far only use information on whether one function value is bigger or smaller than the other or the sign of the derivative. The function value itself in an evaluation point or the value of the derivative has not been used on the decision on how to reduce the interval. Interpolation techniques decide on the location of the iterate x_k based on values in the former iterates.

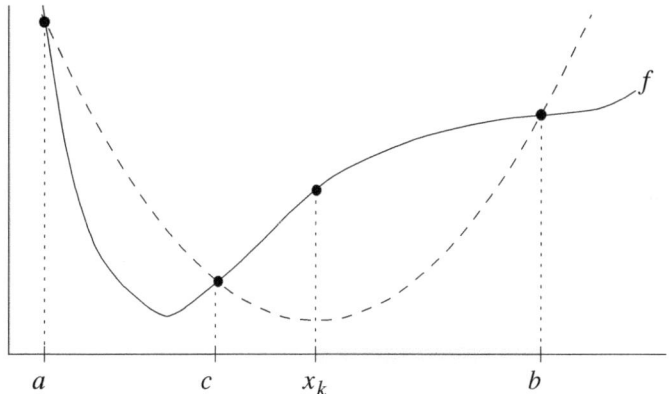

Fig. 5.2. Quadratic interpolation

The central idea of quadratic interpolation is to fit a parabola through the endpoints a, b of the interval and an interior point c and to base the next iterate on its minimum. This works well if

$$f(c) \leq \min\{f(a), f(b)\} \tag{5.3}$$

and the points are not located on one line such that $f(a) = f(b) = f(c)$. It can be shown that the minimum of the corresponding parabola is

---

Algorithm 13 Quadint$([a, b], f, \epsilon)$

Set $k := 1$, $a_1 := a$ and $b_1 := b$
$c := x_0 := \frac{(b+a)}{2}$
Evaluate $f(a_1)$, $f(c) := f(x_0)$, $f(b_1)$
$x_1 := \frac{1}{2} \frac{f(a)(c^2-b^2)+f(c)(b^2-a^2)+f(b)(a^2-c^2)}{f(a)(c-b)+f(c)(b-a)+f(b)(a-c)}$
while $(| c - x_k | > \epsilon)$
    Evaluate $f(x_k)$
    $l := \min\{x_k, x_{k-1}\}$, $r := \max\{x_k, x_{k-1}\}$
    if $(f(r) < f(l))$
        $a_{k+1} := l$, $b_{k+1} := b_k$, $c := r$
    else
        $a_{k+1} := a_k$, $b_{k+1} := r$, $c := l$
    $k := k + 1$
    $x_k := \frac{1}{2} \frac{f(a_k)(c^2-b_k^2)+f(c)(b_k^2-a_k^2)+f(b_k)(a_k^2-c^2)}{f(a_k)(c-b_k)+f(c)(b_k-a_k)+f(b_k)(a_k-c)}$
endwhile

$$x = \frac{1}{2} \frac{f(a)(c^2 - b^2) + f(c)(b^2 - a^2) + f(b)(a^2 - c^2)}{f(a)(c - b) + f(c)(b - a) + f(b)(a - c)}. \qquad (5.4)$$

For use in practice, the algorithm needs many safeguards that switch to Golden Section points if condition (5.3) is not fulfilled. *Brent's method* is doing this in an efficient way; see Brent (1973). We give here only a basic algorithm that works if the conditions are fulfilled.

Example 5.4. Quadratic interpolation is applied to approximate the minimum of $f(x) = x + \frac{16}{x+1}$ with starting interval $[2, 4.5]$ and accuracy $\epsilon = 0.001$. Although the iterate x_k reaches a very good approximation of the minimum point $x^* = 3$ very soon, the proof of convergence is much slower. As can be observed in Table 5.4, the shrinkage of the interval does not have a guaranteed value and is relatively slow. For this reason, the stopping criterion of the algorithm has been put on convergence of the iterate rather than on size of the interval. This example illustrates why it is worthwhile to apply more

Table 5.4. Quadratic interpolation for $f(x) = x + \frac{16}{x+1}$, $[a_0, b_0] = [2, 4.5]$, $\epsilon = 0.001$

k	a_k	b_k	c	x_k	$f(x_k)$
0				3.250	7.0147
1	2.000	4.500	3.250	3.184	7.0081
2	2.000	3.250	3.184	3.050	7.0006
3	2.000	3.184	3.050	3.028	7.0002
4	2.000	3.050	3.028	3.010	7.0000
5	2.000	3.028	3.010	3.005	7.0000
6	2.000	3.010	3.005	3.002	7.0000
7	2.000	3.005	3.002	3.001	7.0000

complex schedules like that of Brent that guarantee a robust reduction to prevent the algorithm from starting to "slice off" parts of the interval.

5.2.5 Cubic interpolation

Cubic interpolation has the same danger of lack of convergence of an enclosing interval, but the theoretical convergence of the iterate is very fast. It has a so-called quadratic convergence. The central idea is to use derivative information in the endpoints of the interval. Together with the function values, x^* is

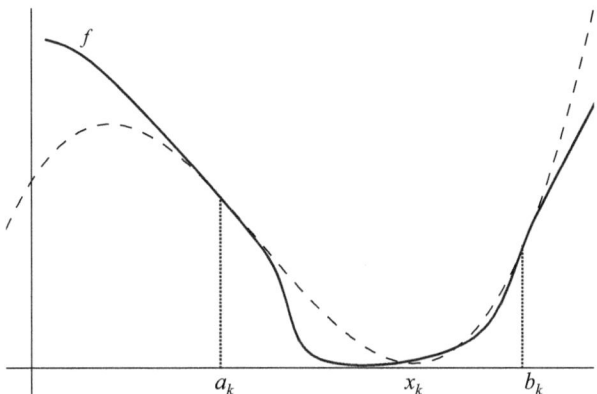

Fig. 5.3. Cubic interpolation

approximated by the minimum of a cubic polynomial. Like in quadratic interpolation, a condition like (5.3) should be checked in order to guarantee that the appropriate minimum locates in the interval $[a, b]$. For cubic interpolation this is

$$f'(a) < 0 \text{ and } f'(b) > 0. \tag{5.5}$$

Given the information $f(a)$, $f'(a)$, $f(b)$ and $f'(b)$ in the endpoints of the interval, the next iterate is given in equation (5.6) in the form that is common in the literature.

$$x_k = b - (b - a)\frac{f'(b) + v - u}{f'(b) - f'(a) + 2v}, \tag{5.6}$$

where $u = f'(a) + f'(b) - 3\frac{f(a)-f(b)}{a-b}$ and $v = \sqrt{u^2 - f'(a)f'(b)}$. The function value and derivative are evaluated in x_k and depending on the sign of the derivative, the interval is reduced to the right or left. Similar to quadratic interpolation, slow reduction of the interval may occur, but on the other hand the iterate converges fast. Notice that the method requires more information, as also the derivatives should be available. The algorithm is sketched without taking safeguards into account with respect to the conditions, or the iterate hitting a stationary point.

Algorithm 14 Cubint$([a,b], f, f', \epsilon)$

Set $k := 1$, $a_1 := a$ and $b_1 := b$

Evaluate $f(a_1)$, $f'(a_1)$, $f(b_1)$, $f'(b_1)$

$u := f'(a) + f'(b) - 3\frac{f(a)-f(b)}{a-b}$, $v := \sqrt{u^2 - f'(a)f'(b)}$

$x_1 := b - (b-a)\frac{f'(b)+v-u}{f'(b)-f'(a)+2v}$

repeat

    Evaluate $f(x_k)$, $f'(x_k)$

    if $f'(x_k) < 0$

        $a_{k+1} := x_k$, $b_{k+1} := b_k$

    else

        $a_{k+1} := a_k$, $b_{k+1} := x_k$

    $k := k + 1$

    $u := f'(a_k) + f'(b_k) - 3\frac{f(a_k)-f(b_k)}{a_k-b_k}$, $v := \sqrt{u^2 - f'(a_k)f'(b_k)}$

    $x_k := b_k - (b_k - a_k)\frac{f'(b_k)+v-u}{f'(b_k)-f'(a_k)+2v}$

until $(|\,x_k - x_{k-1}\,| < \epsilon)$

Example 5.5. Cubic interpolation is applied to find the minimum of $f(x) = x + \frac{16}{x+1}$ with starting interval $[2, 4.5]$ and accuracy $\epsilon = 0.01$. One iteration after reaching the stopping criterion has been given in Table 5.5. For this case, also the interval converges very fast around the minimum point.

Table 5.5. Cubic interpolation for $f(x) = x + \frac{16}{x+1}$, $[a_0, b_0] = [2, 4.5]$, $\epsilon = 0.01$

k	a_k	b_k	x_k	$f(x_k)$	$f'(x_k)$
1	2.000	4.500	3.024	7.0001	0.012
2	2.000	3.024	2.997	7.0000	-0.001
3	2.997	3.024	3.000	7.0000	0.000
4	2.997	3.000	3.000	7.0000	0.000

5.2.6 Method of Newton

In the former examples, the algorithms converge to the minimum point, where the derivative has a value of zero, i.e., it is a stationary point. Methods that look for a point with function value zero can be based on bisection, Brent method, but also on the Newton–Raphson iterative formula: $x_{k+1} = x_k - \frac{f(x_k)}{f'(x_k)}$. If we replace the function f in this formula by its derivative f', we have a basic method for looking for a stationary point. We have already seen in the elaboration in Chapter 4 that the method may converge to a minimum, maximum or infliction point.

In order to converge to a minimum point, in principle the second-order derivative of an iterate should be positive, i.e., $f''(x_k) > 0$. If we have a starting interval, also safeguards should be included in the algorithm to prevent

---

Algorithm 15 Newt(x_0, f, ϵ)

Set $k := 0$
repeat

$\quad x_{k+1} := x_k - \frac{f'(x_k)}{f''(x_k)}$
$\quad k := k + 1$
until $(\mid x_k - x_{k-1} \mid < \epsilon)$

---

the iterates from leaving the interval. The basic shape of the method without any safeguards is given in Algorithm 15.

Example 5.6. The method of Newton is used for the example function $f(x) = x + \frac{16}{x+1}$ with starting point $x_0 = 2$ and accuracy $\epsilon = 0.01$. Theoretically the method of Newton has the same convergence rate as cubic interpolation. For this specific example one can observe a similar speed of convergence.

Table 5.6. Newton for $f(x) = x + \frac{16}{x+1}$, $x_0 = 2$, $\epsilon = 0.01$

k	x_k	$f(x_k)$	$f'(x_k)$	$f''(x_k)$
0	2.000	7.3333	-0.778	1.185
1	2.656	7.0323	-0.197	0.655
2	2.957	7.0005	-0.022	0.516
3	2.999	7.0000	0.000	0.500
4	3.000	7.0000	0.000	0.500

5.3 Algorithms not using derivative information

In Section 5.2, we have seen that several methods use derivative information and others do not. Let us consider methods for finding optima of functions of several variables, $f : \mathbb{R}^n \to \mathbb{R}$. When derivative information is not available, or one does not want to use it, there are several options to be considered. One approach often used is to apply methods that use derivative information and to approximate the derivative in each iteration numerically. Another option is to base the search directions in Algorithm 9 on directions that are determined by only using the values of the function evaluations. A last option is the use of so-called direct search methods.

From this last class, we will describe the so-called Downhill Simplex method due to Nelder and Mead (1965). It is popular due to its attractive geometric description and robustness and also its appearance in standard software like MATLAB (www.mathworks.com) and the Numerical Recipes of Press et al. (1992). It will be described in Section 5.3.1. Press et al. (1992) also mention that "Powell's method is almost surely faster in all likely applications." The method of Powell is based on generating search directions built on earlier directions like in Algorithm 9. It is described in Section 5.3.2.

5.3.1 Method of Nelder and Mead

Like in evolutionary algorithms (see Davis, 1991, and Section 7.5), the method works with a set of points that is iteratively updated. The iterative set $P = \{p_0, \ldots, p_n\}$ is called a simplex, because it contains $n + 1$ points in an n-dimensional space. The term *Simplex method* used by Nelder and Mead (1965) should not be confused with the Simplex method for Linear Optimization. Therefore, it is also called the *Polytope method* to distinguish the two. The initial set of points can be based on a starting point x_0 by taking $p_0 = x_0$, $p_i = x_0 + \delta e_i$, $i = 1, \ldots, n$, where δ is a scaling factor and e_i the ith unit vector. The following ingredients are important in the algorithm and define the trial points.

- The two worst points $p_{(n)} = \mathrm{argmax}_{p \in P} f(p)$, $p_{(n-1)} = \mathrm{argmax}_{p \in P \setminus p_{(n)}} f(p)$ in P and lowest point $p_{(0)} = \mathrm{argmin}_{p \in P} f(p)$ are identified.
- The centroid c of all but the highest point is used as building block

$$c = \frac{1}{n} \sum_{i \neq (n)} p_i. \tag{5.7}$$

Algorithm 16 NelderMead(x_0, f, ϵ)

Set $k := 0$, $P := \{p_0, \ldots, p_n\}$ with $p_0 := x_0$ and $p_i := x_0 + \delta e_i$ $i = 1, \ldots, n$
Evaluate $f(p_i)$ $i = 1, \ldots, n$
Determine points $p_{(n)}, p_{(n-1)}$ and $p_{(0)}$ in P
with corresponding values $f_{(n)}, f_{(n-1)}$ and $f_{(0)}$
while $(f_{(n)} - f_{(0)} > \epsilon)$
$\qquad c := \frac{1}{n} \sum_{i \neq (n)} p_i$
$\qquad x^{(r)} := c + (c - p_{(n)})$, evaluate $f(x^{(r)})$
$\qquad$**if** $(f_{(0)} < f(x^{(r)}) < f_{(n-1)})$
$\qquad\qquad P := P \setminus \{p_{(n)}\} \cup \{x^{(r)}\}$ $\qquad\qquad\qquad\qquad$ $x^{(r)}$ replaces $p_{(n)}$ in P
$\qquad$**if** $(f(x^{(r)}) < f_{(0)})$
$\qquad\qquad x^{(e)} := c + 1.5(c - p_{(n)})$, evaluate $f(x^{(e)})$
$\qquad\qquad P := P \setminus \{p_{(n)}\} \cup \{\mathrm{argmin}\{f(x^{(e)}), f(x^{(r)})\}\}$ $\qquad$ best trial replaces $p_{(n)}$
$\qquad$**if** $(f(x^{(r)}) \geq f_{(n-1)})$
$\qquad\qquad x^{(c)} := c + 0.5(c - p_{(n)})$, evaluate $f(x^{(c)})$
$\qquad\qquad$**if** $(f(x^{(c)}) < f(x^{(r)}) < f_{(n)})$
$\qquad\qquad\qquad P := P \setminus \{p_{(n)}\} \cup \{x^{(c)}\}$ $\qquad\qquad\qquad\qquad$ replace $p_{(n)}$ by $x^{(c)}$
$\qquad\qquad$**else**
$\qquad\qquad\qquad$**if** $(f(x^{(c)}) > f(x^{(r)}))$
$\qquad\qquad\qquad\qquad P := P \setminus \{p_{(n)}\} \cup \{x^{(r)}\}$
$\qquad\qquad\qquad$**else**
$\qquad\qquad\qquad\qquad p_i := \frac{1}{2}(p_i + p_{(0)}), i = 0, \ldots, n$ $\qquad\qquad$ full contraction
$\qquad\qquad\qquad\qquad$Evaluate $f(p_i), i = 1, \ldots, n$
$\qquad\qquad\qquad\qquad P := \{p_0, \ldots, p_n\}$
$\qquad k := k + 1$
endwhile

- A trial point is based on *reflection* step: $x^{(r)} = c + (c - p_{(n)})$, Figure 5.4(a).
- When the former step is successful, a trial point is based on an *expansion* step $x^{(e)} = c + 1.5(c - p_{(n)})$, shown in Figure 5.4(c).
- In some cases a *contraction* trial point is generated as shown in Figure 5.4(b); $x^{(c)} = c + 0.5(c - p_{(n)})$.
- If the trials are not promising, the simplex is shrunk via a so-called *multiple contraction* toward the point with lowest value $p_i := \frac{1}{2}(p_i + p_{(0)}), i = 0, \ldots, n$.

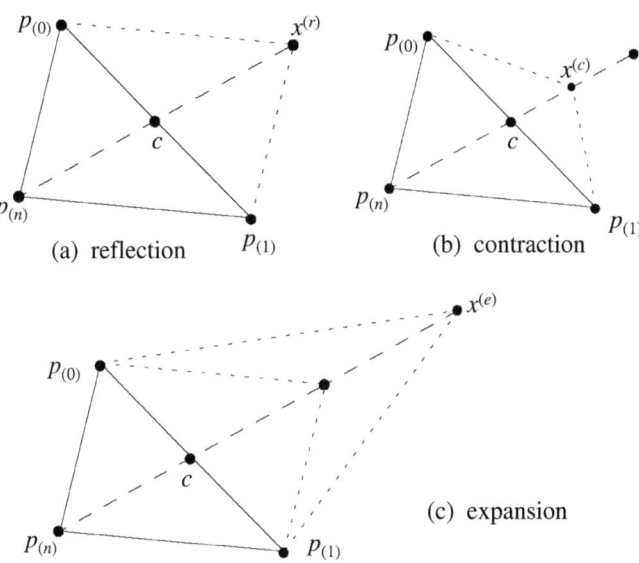

(a) reflection   (b) contraction

(c) expansion

Fig. 5.4. Basic steps of the Nelder and Mead algorithm

In the description we fix the size of reflection, expansion and contraction. Usually this depends on parameters with its value depending on the dimension of the problem. A complete description is given in Algorithm 16.

Example 5.7. Consider the function $f(x) = 2x_1^2 + x_2^2 - 2x_1x_2 + |x_1 - 3| + |x_2 - 2|$. Let the initial simplex be given by $p_0 = (1, 2)^T$, $p_1 = (1, 0)^T$ and $p_2 = (2, 1)^T$. The first steps are depicted in Figure 5.5. We can see at part (a) that first a reflection step is taken, the new point becomes $p_{(1)}$. However, at the next iteration, the reflection point satisfies neither condition $f_{(0)} < f(x^{(r)}) < f_{(n-1)}$ nor $f(x^{(r)}) < f_{(0)}$, thus the contraction point is calculated (see Figure 5.5(b)). As it has a better function value than $f(x^{(r)})$, $p_{(n)}$ is replaced by this point. We can also see that $f(x^{(c)}) < f_{(n-1)}$ as the ordering changes in Figure 5.5(c). One can observe that when the optimum seems to be inside the polytope, its size decreases leading toward fulfillment

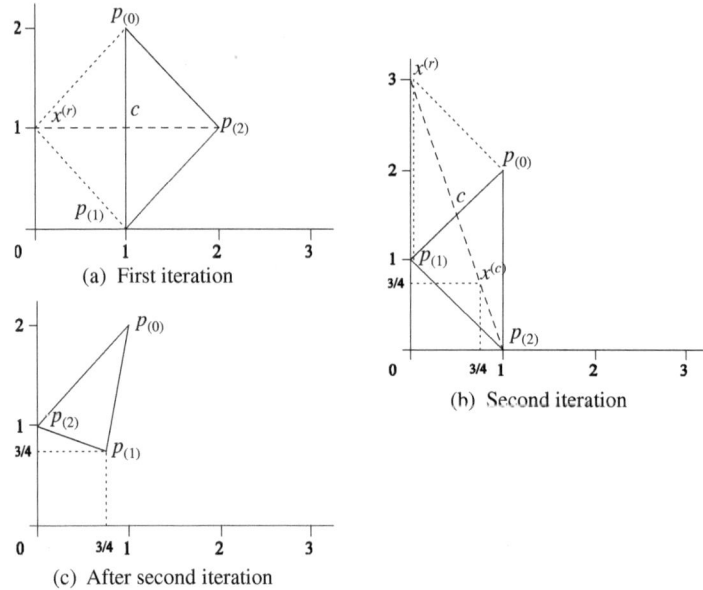

Fig. 5.5. Nelder and Mead method at work

of the termination condition. The FMINSEARCH algorithm in MATLAB is an implementation of Nelder–Mead. From a starting point $p_0 = x_0$ a first small simplex is built. Running the algorithm with default parameter values and $x_0 = (1, 0)^T$ requires 162 function evaluations before stopping criteria are met. The evaluated sample points are depicted in Figure 5.6.

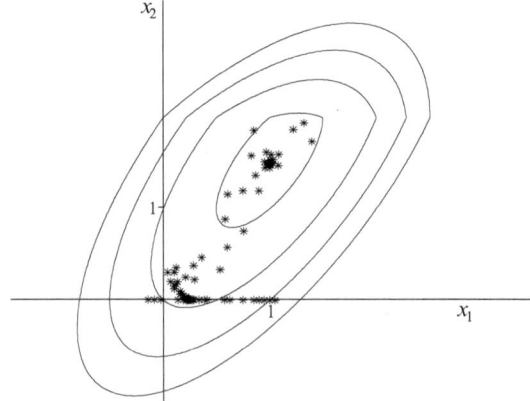

Fig. 5.6. Points generated by NelderMead on $f(x) = 2x_1^2 + x_2^2 - 2x_1x_2 + |x_1 - 3| + |x_2 - 2|$. FMINSEARCH with default parameter values

5.3.2 Method of Powell

In this method, credited to Powell (1964), a set of directions $(d_1, \ldots, d_n)$ is iteratively updated to approximate the direction pointing to x^*. An initial point x_0 is given, that will be named $x_1^{(1)}$. At each iteration k, n steps are taken using the n directions. In each step, $x_{i+1}^{(k)} = x_i^{(k)} + \lambda d_i$, where the step size λ is supposed to be optimal, i.e., $\lambda = \operatorname{argmin}_\mu f(x_i^{(k)} + \mu d_i)$. The direction set is initialized with the coordinate directions, i.e., $(d_1, \ldots, d_n) = (e_1, \ldots, e_n)$. In fact the first iteration works as the so-called Cyclic Coordinate Method. However, in the method of Powell (see Algorithm 17) instead of starting over with the same directions, they are updated as follows. Direction

---

Algorithm 17 Powell(x_0, f, ϵ)

Set $k := 0$, $(d_0, \ldots, d_n) := (e_0, \ldots, e_n)$, and $x_1^{(1)} := x_0$
repeat
    $k := k + 1$
    for $(i = 1, \ldots, n)$ **do**
        Determine step size $\lambda := \operatorname{argmin}_\mu f(x_i^{(k)} + \mu d_i)$
        $x_{i+1}^{(k)} := x_i^{(k)} + \lambda d_i$
    $d := x_{n+1}^{(k)} - x_1^{(k)}$
    $x_1^{(k+1)} := x_{n+1}^{(k)} + \lambda d$ where $\lambda := \operatorname{argmin}_\mu f(x_{n+1}^{(k)} + \mu d)$
    $d_i := d_{i+1}, i = 1, \ldots, n-1, d_n := d$
until $(|f(x_1^{(k+1)}) - f(x_1^{(k)})| < \epsilon)$

$d = x_{n+1}^{(k)} - x_1^{(k)}$ is the overall direction in the kth iteration. Let the starting point for the next iteration be in that direction: $x_1^{(k+1)} = x_{n+1}^{(k)} + \lambda d$ with optimal step size λ. The old directions are shifted, $d_i = d_{i+1}, i = 1, \ldots, n-1$, and the last one is our approximation, $d_n = d$. The iterations continue with the updated directions until $|f(x_1^{(k+1)}) - f(x_1^{(k)})| < \epsilon$.

Example 5.8. Consider the function $f(x) = 2x_1^2 + x_2^2 - 2x_1 x_2 + |x_1 - 3| + |x_2 - 2|$ and let $x_0 = (0,0)^T$. The steps of the method of Powell are shown in Figure 5.7. Observe that points $x_1^{(1)}, x_3^{(1)}, x_1^{(2)}$ and $x_1^{(2)}, x_3^{(2)}, x_1^{(3)}$ lie on a common line, that has the direction d of the corresponding iteration. In this example, the optimum is found after only three iterations. Notice that in each step an exact line search is done in order to obtain the optimal step length λ.

In both the Polytope method and the method of Powell the direction of the new step depends on the last n points. This is necessary to generate a descent direction when only function values are known. In the next sections we will see that derivative information gives easier access to descent directions.

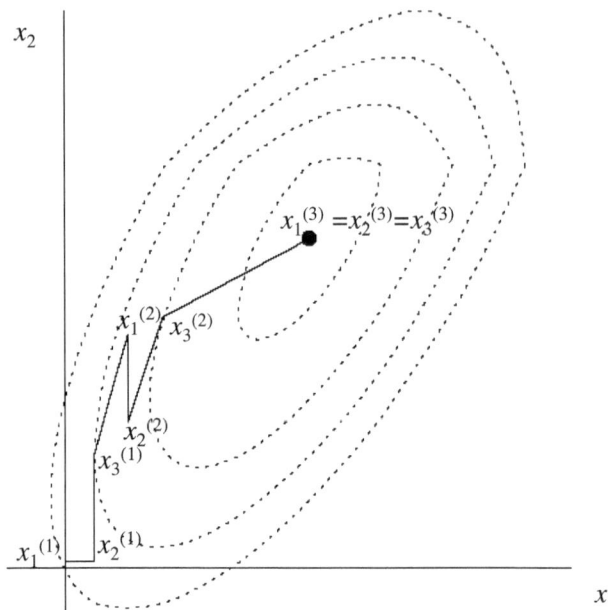

Fig. 5.7. Example run of the Powell method

5.4 Algorithms using derivative information

When the function to be minimized is continuously differentiable, i.e., $f : \mathbb{R}^n \to \mathbb{R} \in C^1$, methods using derivative information are likely to be more efficient. Some methods may even use Hessean information if that is available. These methods usually can be described by the general scheme of descent direction methods introduced in Algorithm 9. There are two crucial points in these algorithms: the choice of the descent direction and the size of the step to take. The methods are usually named after the way the descent direction is defined, and they have different versions and modifications depending on how the step length is chosen.

The first method we discuss is the Steepest descent algorithm in Section 5.4.1, where the steepest direction is chosen based on the first-order Taylor expansion. As a second algorithm, the Newton method is explained in Section 5.4.2. It is based on the second-order Taylor expansion and uses second derivative information. These two methods are based on local information only, i.e., the Taylor expansion of the function at the given point. Conjugate gradient and Quasi-Newton methods also use information from previous steps to improve the next direction. These advanced methods are introduced in Section 5.4.3 and 5.4.4, respectively. Finally, we discuss the consequence of using practical line search methods together with the concept of trust region methods in Section 5.4.5.

5.4.1 Steepest descent method

This method is quite historical in the sense that it was introduced in the middle of the 19th century by Cauchy. The idea of the method is to decrease the function value as much as possible in order to reach the minimum early. Thus, the question is in which direction the function decreases most. The first-order Taylor expansion of f near point x in the direction r is

$$f(x + r) \approx f(x) + \nabla f(x)^T r.$$

So, we search for the direction

$$\min_{r \in \mathbb{R}^n} \frac{\nabla f(x)^T r}{\|r\|},$$

which is for the Euclidean norm the negative gradient, i.e., $r = -\nabla f(x)$ (see Figure 5.8). That is why this method is also called the *gradient method*.

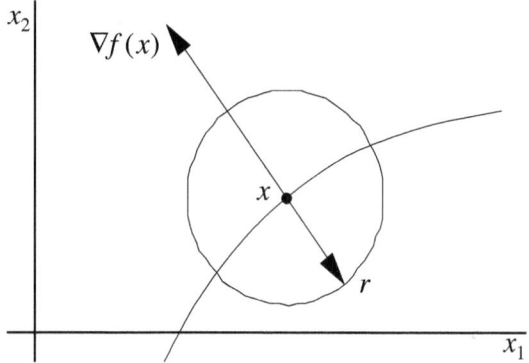

Fig. 5.8. Steepest descent direction

In Figure 5.9 we can see an example run of the method, when the optimal step length is taken for a quadratic function. Notice that the steps are perpendicular. This is not a coincidence. When the step length is optimal at the new point, the derivative is zero in the last direction. The new direction can only be perpendicular. This is called the zigzag effect, and it makes the convergence slow when the optimum is near.

Example 5.9. Let $f(x) = (x_1 - 3)^2 + 3(x_2 - 1)^2 + 2$ and $x_0 = (0,0)^T$. The gradient is $\nabla f(x) = \begin{pmatrix} 2(x_1 - 3) \\ 6(x_2 - 1) \end{pmatrix}$, the steepest descent $-\nabla f(x_0) = \begin{pmatrix} 6 \\ 6 \end{pmatrix}$. We take as first search direction $r_0 = (1,1)^T$. The optimum step size λ can be found by minimizing $\varphi_{r_0}(\mu) = f(x_0 + \mu r_0)$ over μ. For a quadratic function we can consider finding the stationary point, such that

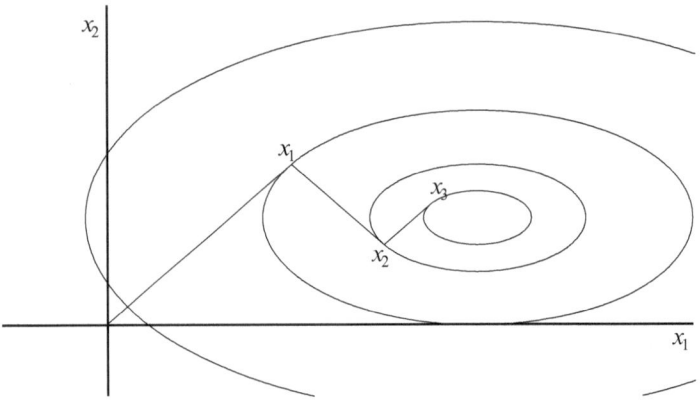

Fig. 5.9. Example run of steepest descent method

$$\varphi'(\lambda) = r_0^T \nabla f(x_0 + \lambda r_0) = (1,1)^T \begin{pmatrix} 2(x_1 - 3) \\ 6(x_2 - 1) \end{pmatrix} = 2(\lambda - 3) + 6(\lambda - 1) = 0.$$

This gives the optimal step size of $\lambda = \frac{3}{2}$. The next iterate is $x_1 = (x_0 + \lambda r_0) = (0,0)^T + \frac{3}{2}(1,1)^T = (1.5, 1.5)^T$. Following the steepest descent process where we keep the same length of the search vector leads to the iterates in Table 5.7. Notice that $\|\nabla f_k\|$ is getting smaller, as x_k is converging to the minimum point. Moreover, notice that $r_k^T r_{k-1} = 0$.

Table 5.7. Steepest descent iterations, $f(x) = (x_1 - 3)^2 + 3(x_2 - 1)^2 + 2$ and $x_0 = 0$

k	$-\nabla f_{k-1}^T$	r_{k-1}^T	λ	x_k^T	$f(x_k)$
0				(0,0)	12
1	(6,6)	(1,1)	$\frac{3}{2}$	(1.5, 1.5)	5
2	(3,-3)	(1,-1)	$\frac{3}{4}$	(2.25, 0.75)	2.75
3	(1.5, 1.5)	(1,1)	$\frac{3}{8}$	(2.625, 1.125)	2.1875

    In practical implementations, computing the optimal step length far away from x^* can be unnecessary and time consuming. Therefore, fast inexact line search methods have been suggested to approximate the optimal step length. We discuss these approaches in Section 5.4.5.

5.4.2 Newton method

We have already seen the Newton method in the univariate case in Section 5.2.6. For multivariate optimization the generalization is straightforward:

$$x_{k+1} = x_k - H_f^{-1}(x_k)\nabla f(x_k).$$

But where does this formula come from? Let us approximate the function f with its second-order Taylor expansion

$$T(x+r) = f(x) + \nabla f(x)^T r + \frac{1}{2}r^T H_f(x)r.$$

Finding the minimum of $T(x+r)$ in r can give us a new direction towards x^*. Having a positive definite Hessean H_f (see Section 3.3), the minimum is the solution of $\nabla T(x+r) = 0$. Thus, we want to solve the linear equation system

$$\nabla T(x+r) = \nabla f(x) + H_f(x)r = 0$$

in r. Its solution $r = -H_f^{-1}(x)\nabla f(x)$ gives direction as well as step size.

The above construction ensures that for quadratic functions the optimum (if it exists) is found in one step.

Example 5.10. Consider the same minimization problem as in Example 5.9, i.e., minimize $f(x) = (x_1 - 3)^2 + 3(x_2 - 1)^2 + 2$ with starting point $x_0 = 0$. Gradient $\nabla f(x) = \begin{pmatrix} 2(x_1 - 3) \\ 6(x_2 - 1) \end{pmatrix}$ while the Hessean $H_f(x) = \begin{pmatrix} 2 & 0 \\ 0 & 6 \end{pmatrix}$. Thus,

$$x_1 = x_0 - H_f^{-1}\nabla f(x_0) = \begin{pmatrix} 0 \\ 0 \end{pmatrix} - \begin{pmatrix} 1/2 & 0 \\ 0 & 1/6 \end{pmatrix}\begin{pmatrix} -6 \\ -6 \end{pmatrix} = \begin{pmatrix} 3 \\ 1 \end{pmatrix}.$$ At x_1 the

gradient is zero, the Hessean is positive definite, thus we have reached the optimum.

5.4.3 Conjugate gradient method

This class of methods can be viewed as a modification of the steepest descent method, where in order to avoid the zigzagging effect, at each iteration the direction is modified by a combination of the earlier directions:

$$r_k = -\nabla f_k + \beta_k r_{k-1}. \tag{5.8}$$

These corrections ensure that $r_1, r_2, \ldots, r_n$ are so-called conjugate directions. This means that there exists a matrix A such that $r_i^T A r_j = 0$, $\forall i \neq j$. For instance, the coordinate directions (the unit vectors) are conjugate. Just take A as the unit matrix. The underlying idea is that A is the inverse of the Hessean. One can derive that using exact line search the optimum is reached in at most n steps for quadratic functions.

Having the direction r_k, the next iterate is calculated in the usual way:

$$x_{k+1} = x_k + \lambda r_k$$

where λ is the optimal step length $\operatorname{argmin}_\mu f(x_k + \mu r_k)$, or its approximation.

The parameter β_k can be calculated using different formulas. Hestenes and Stiefel (1952) suggested

$$\beta_k = \frac{\nabla f_k^T (\nabla f_k - \nabla f_{k-1})}{r_k^T (\nabla f_k - \nabla f_{k-1})}. \tag{5.9}$$

Later, Fletcher and Reeves (1964) examined

$$\beta_k = \frac{\|\nabla f_k\|^2}{\|\nabla f_{k-1}\|^2}, \tag{5.10}$$

and lastly the formula of Polak and Ribière (1969) is

$$\beta_k = \frac{\nabla f_k^T (\nabla f_k - \nabla f_{k-1})}{\|\nabla f_{k-1}\|^2}. \tag{5.11}$$

These formulas are based on the quadratic case where $f(x) = \frac{1}{2}x^T A x + b^T x + c$ for a positive definite A. For this function, the aim is to have A-conjugate directions, so $r_j^T A r_i, \ \forall j \neq i$. Plugging (5.8) into $r_k^T A r_{k-1} = 0$ gives $-\nabla f_k^T A r_{k-1} + \beta_k r_{k-1}^T A r_{k-1} = 0$ such that

$$\beta_k = \frac{\nabla f_k^T A r_{k-1}}{r_{k-1}^T A r_{k-1}}.$$

Now, having $\nabla f(x) = Ax + b$ gives $\nabla f(x_k) = A(x_{k-1} + \lambda r_{k-1}) + b = \nabla f_{k-1} + \lambda A r_{k-1}$ such that $\nabla f_k - \nabla f_{k-1} = \lambda A r_{k-1}$. Thus,

$$\beta_k = \frac{\nabla f_k^T A r_{k-1}}{r_{k-1}^T A r_{k-1}} = \frac{\nabla f_k^T (\nabla f_k - \nabla f_{k-1})}{r_k^T (\nabla f_k - \nabla f_{k-1})}.$$

This is exactly the formula of Hestenes and Stiefel. In fact, for the quadratic case all three formulas are equal, and the optimum is found in at most n steps.

Example 5.11. Consider the instance of Example 5.9 with $f(x) = (x_1 - 3)^2 + 3(x_2 - 1)^2 + 2$ and $x_0 = (0,0)^T$. In the first iteration, we follow the steepest descent, such that $\nabla f(x_0) = \begin{pmatrix} -6 \\ -6 \end{pmatrix}$ gives our choice $r_0 = (1,1)^T$, $\lambda = \frac{3}{2}$ and $x_1 = (1.5, 1.5)^T$. Now we follow the conjugate direction given by (5.8) and Fletcher–Reeves (5.10). Given that $\nabla f(x_1) = (-3,3)^T$, $\|\nabla f(x_0)\|^2 = 72$ and $\|\nabla f(x_1)\|^2 = 18$, the next direction is determined by

$$r_1 = -\nabla f_1 + \beta_1 r_0 = -\nabla f_1 + \frac{\|\nabla f_1\|^2}{\|\nabla f_0\|^2} r_0 = \begin{pmatrix} 3 \\ -3 \end{pmatrix} + \frac{18}{72} \begin{pmatrix} 6 \\ 6 \end{pmatrix} = \begin{pmatrix} 4.5 \\ -1.5 \end{pmatrix}.$$

This direction points directly to the minimum point $x^* = (3,1)^T$, see Figure 5.10. Notice that r_0 and r_1 are conjugate with respect to the Hessean H of f:

$$r_0^T H r_1 = (1,1) \begin{pmatrix} 2 & 0 \\ 0 & 6 \end{pmatrix} \begin{pmatrix} 4.5 \\ -1.5 \end{pmatrix} = 0.$$

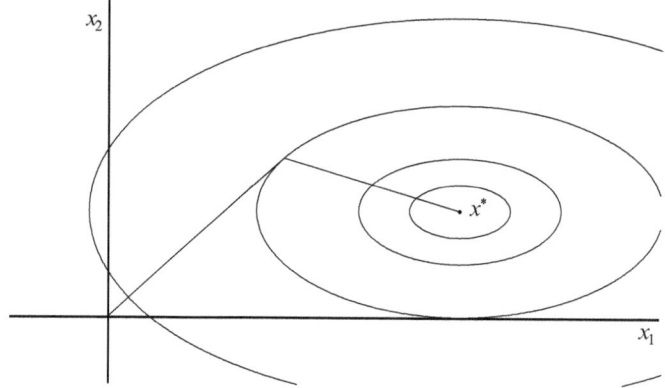

Fig. 5.10. Example run of conjugate gradient method

5.4.4 Quasi-Newton method

The name tells us that these methods work similarly as the Newton method. The main idea is to approximate the Hessean matrix instead of computing it at every iteration. Recall that the Newton method computes the search direction as

$$r_k = -H_f(x_k)^{-1}\nabla f(x_k),$$

where $H_f(x_k)$ should be positive definite. In order to avoid problems with non-positive-definite or noninvertible Hessean matrices and in addition to save Hessean evaluation, quasi-Newton methods approximate $H_f(x_k)$ by B_k using an updating formula $B_{k+1} = B_k + U_k$.

The updating should be such that at each step the new curvature information is built in the approximated Hessean. Using the second-order Taylor expansion of function f,

$$T(x_k + r) \approx f(x_k) + \nabla f(x_k)^T r + \frac{1}{2} r^T H_f(x_k) r,$$

one can obtain that

$$\nabla f(x_k + r) \approx \nabla T(x_k + r) = \nabla f(x_k) + H_f(x_k) r.$$

Taking $r = r_k$ and denoting $y_k = \nabla f(x_{k+1}) - \nabla f(x_k)$ gives

$$y_k \approx H_f(x_k) r_k. \tag{5.12}$$

Equation (5.12) gives the so-called *quasi-Newton condition*, that is, $y_k = B_k r_k$ must hold for every B_k and each search direction $r_k = x_{k+1} - x_k$ we take. Apart from (5.12), we also require B_k to be positive definite and symmetric, although that is not necessary.

For a rank one update, that is, $B_{k+1} = B_k + \alpha_k u_k u_k^T (u_k \in \mathbb{R}^n)$, the above requirements define the update:

$$B_{k+1} = B_k + \frac{1}{(y_k - B_k r_k)^T r_k}(y_k - B_k r_k)(y_k - B_k r_k)^T. \tag{5.13}$$

This is called the *symmetric rank one formula* (SR1).

In general, after updating the approximate Hessean matrix, its inverse should be computed to obtain the direction. Fortunately, using the Sherman–Morrison formula we can directly update the inverse matrix. For SR1 formula (5.13), denoting $M_k = B_k^{-1}$,

$$M_{k+1} = M_k + \frac{1}{(r_k - M_k y_k)^T y_k}(r_k - M_k y_k)(r_k - M_k y_k)^T.$$

Two popular rank two update formulas deserve to be mentioned. The general form for rank two formulas is $B_{k+1} = B_k + \alpha_k u_k u_k^T + \beta_k v_k v_k^T$. One of them is the *Davidon–Fletcher–Powell formula* (DFP), that determines B_{k+1} or M_{k+1} as

$$B_{k+1} = B_k + \frac{(y_k - B_k r_k)(y_k - B_k r_k)^T}{y_k^T r_k} - \frac{B_k r_k r_k^T B_k}{y_k^T r_k} + \frac{r_k^T B_k r_k y_k y_k^T}{(y_k^T r_k)^2}$$

$$M_{k+1} = M_k + \frac{r_k r_k^T}{y_k^T r_k} - \frac{M_k y_k y_k^T M_k}{y_k^T M_k y_k}. \tag{5.14}$$

Later, the Broyden–Fletcher–Goldfarb–Shanno (BFGS) method was discovered by Broyden, Fletcher, Goldfarb, and Shanno independently of each other around 1970. Nowadays mostly this update formula is used. The updating formulas are

$$B_{k+1} = B_k + \frac{y_k y_k^T}{y_k^T r_k} - \frac{B_k r_k r_k^T B_k}{r_k^T B_k r_k}$$

$$M_{k+1} = M_k + \frac{(r_k - M_k y_k)(r_k - M_k y_k)^T}{y_k^T r_k} - \frac{M_k y_k y_k^T M_k}{y_k^T r_k} + \frac{y_k^T M_k y_k r_k r_k^T}{(y_k^T r_k)^2}.$$

Example 5.12. We now elaborate the DFP method based on the instance of Example 5.9 with $f(x) = (x_1 - 3)^2 + 3(x_2 - 1)^2 + 2$ and $x_0 = (0,0)^T$. In the first iteration, we follow the steepest descent. $\nabla f(x_0) = \begin{pmatrix} -6 \\ -6 \end{pmatrix}$ and exact line search gives $x_1 = (1.5, 1.5)^T$. In terms of the quasi-Newton concept, direction $r_0 = x_1 - x_0 = (1.5, 1.5)^T$ and $y_0 = \nabla f_1 - \nabla f_0 = (3,9)^T$. Now we can determine all ingredients to compute the updated matrix of (5.14). Keeping in mind that M_0 is the unit matrix, such that $M_0 y_0 = y_0$,

$$r_0 r_0^T = \frac{9}{4}\begin{pmatrix} 1 & 1 \\ 1 & 1 \end{pmatrix}, M_0 y_0 y_0^T M_0 = 9\begin{pmatrix} 1 & 3 \\ 3 & 9 \end{pmatrix}, r_0^T y_0 = 18 \text{ and } y_0^T M_0 y_0 = 90.$$

The updated multiplication matrix M_1 is now determined by (5.14):

$$M_1 = \begin{pmatrix} 1 & 0 \\ 0 & 1 \end{pmatrix} + \frac{1}{8} \begin{pmatrix} 1 & 1 \\ 1 & 1 \end{pmatrix} - \frac{1}{10} \begin{pmatrix} 1 & 3 \\ 3 & 9 \end{pmatrix} = \frac{1}{40} \begin{pmatrix} 41 & -7 \\ -7 & 9 \end{pmatrix}.$$

Notice that M_1 fulfills the (inverse) quasi-Newton condition $r_0 = M_1 y_0$. Now we can determine the search direction

$$r_1 = -M_1 \nabla f_1 = \frac{1}{40} \begin{pmatrix} 41 & -7 \\ -7 & 9 \end{pmatrix} \begin{pmatrix} 3 \\ -3 \end{pmatrix} = \frac{6}{5} \begin{pmatrix} 3 \\ -1 \end{pmatrix}.$$

This is the same direction of search as found by the conjugate direction method in Example 5.11 and points to the minimum point $x^* = (3, 1)^T$. Further determination of M_2 is more cumbersome by hand, although easy with a matrix manipulation program. One can verify that $M_2 = H_f^{-1}$ as should be the case for quadratic functions.

5.4.5 Inexact line search

In almost all descent direction methods, a line search is done in each step. So far we have only used the optimal step length, which means that exact line search was supposed. We have already seen that for quadratic functions the optimal step length is easy to compute. Otherwise a one-dimensional optimization method (see Section 5.2) can be used. When we are still far away from the minimum, computing a very good approximation of the optimal step length is not efficient usually. But how to know that we are still far away from the optimum and that an approximation is good enough? Of course there is no exact answer to these questions, but some rules can be applied. For instance, we suspect that $\|\nabla f(x)\| \to 0$ as $x \to x^*$. To avoid a too big or too small step, a sufficient decrease in objective is required. For a small $0 < \alpha < 1$,

$$f_k + (1 - \alpha)\lambda \nabla f_k^T r_k < f(x_k + \lambda r_k) \tag{5.15}$$
$$f(x_k + \lambda r_k) < f_k + \alpha \lambda \nabla f_k^T r_k \tag{5.16}$$

must hold. Denoting $\varphi_{r_k}(\lambda) = f(x_k + \lambda r_k)$ we can write (5.15)–(5.16) together as

$$\varphi_{r_k}(0) + (1 - \alpha)\varphi'_{r_k}(0)\lambda < \varphi_{r_k}(\lambda) < \varphi_{r_k}(0) + \alpha \varphi'_{r_k}(0)\lambda.$$

(5.15)-(5.16) is called the Goldstein condition. Inequality (5.16) alone is called the Armijo condition. The idea is depicted in Figure 5.11. Inequality (5.15) states that λ has to be greater than a lower bound $\underline{\lambda}$. The Armijo condition (5.16) gives an upper bound $\overline{\lambda}$ on the step size. We can have more disconnected intervals for λ, and (5.15) may exclude the optimal solution, as it does exclude a local optimum in Figure 5.11.

To avoid this exclusion, one can use the Wolfe condition. That condition says that the derivative in the new point has to be smaller than in the old point; for a parameter $0 < \sigma < 1$,

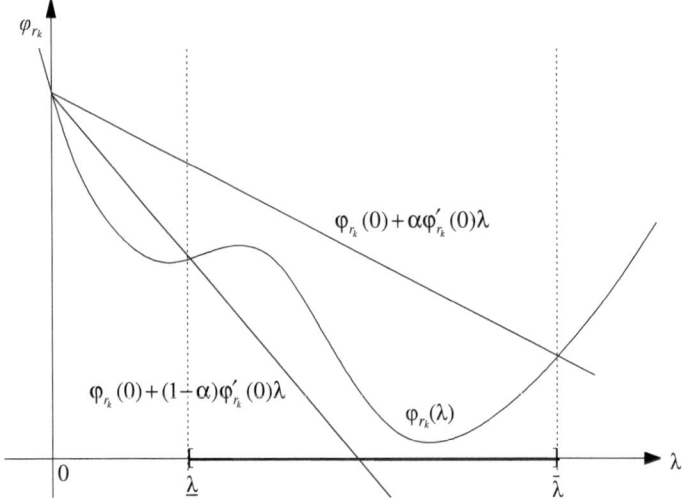

Fig. 5.11. Goldstein condition

$$\varphi'_{r_k}(\lambda) < \sigma \varphi'_{r_k}(0), \tag{5.17}$$

or alternatively

$$\nabla f(x_k + \lambda r_k)^T r_k < \sigma \nabla f(x_k)^T r_k.$$

The Wolfe condition (5.17) together with the Armijo condition (5.16) is called the Wolfe conditions. In the illustration, (5.16) and (5.17) mean that step size λ must belong to one of the intervals shown in Figure 5.12.

The good news about these conditions is that the used line search can be very rough. If the step length fulfills these conditions, then convergence can be proved.

In practice, usually a backtracking line search is done until the chosen conditions are fulfilled. The concept of backtracking line search is very easy. Given a (possibly large) initial step length λ_0, decrease it proportionally with a factor $0 < \beta < 1$ until the chosen condition is fulfilled (see Algorithm 18).

Algorithm 18 BacktrackLineSearch($\lambda_0, \varphi_{r_k}, \beta$)

$k := 1$
while (conditions not fulfilled)
$\quad \lambda_k := \beta \lambda_{k-1}$
$\quad k := k + 1$
endwhile

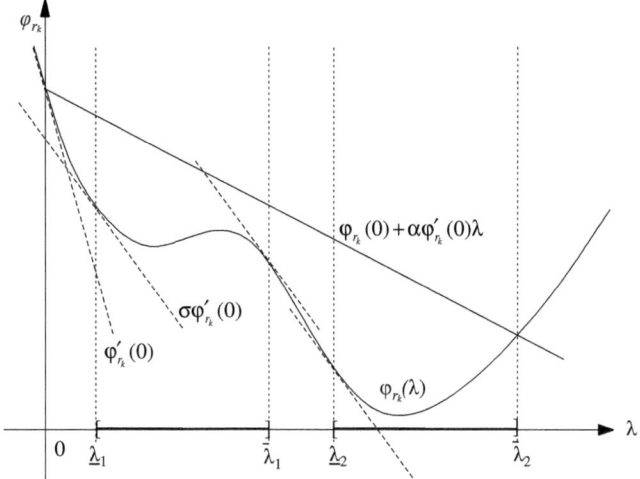

Fig. 5.12. Wolfe conditions

5.4.6 Trust region methods

Trust region methods have a different concept than general descent methods. The idea is first to decide the step size, and then to optimize for the best direction. The step size defines the radius Δ of the trust region, where the approximate function (usually the second-order Taylor expansion) is trusted to behave similarly as the original function. Within radius Δ (or maximum step size) the best direction is calculated according to the approximate function $m_k(x)$, i.e.,

$$\min_{\|r\|<\Delta} m_k(x_k + r), \tag{5.18}$$

where usually

$$m_k(x_k + r) = f(x_k) + \nabla f(x_k)^T r + \frac{1}{2} r^T H_f(x_k) r.$$

To control that we are doing well, the adequacy of the trust radius is checked. Hence, the predicted reduction $m_k(x_k) - m_k(x_k + r_k)$ and the actual reduction $f(x_k) - f(x_k + r_k)$ are compared. For a given parameter μ, if

$$\left(\rho_k = \frac{f(x_k) - f(x_k + r_k)}{m_k(x_k) - m_k(x_k + r_k)} \right) > \mu \tag{5.19}$$

holds, the trust region and the step are accepted. Otherwise the radius is reduced and the direction is optimized again, see Figure 5.13. When the prediction works very well, we can increase the trust region. Given a second parameter $\nu > \mu$, if

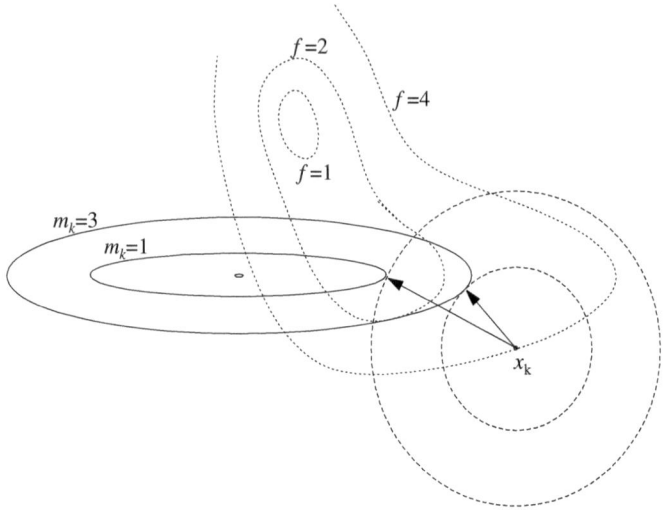

Fig. 5.13. For different trust radius different directions are optimal

$$\rho_k > \nu,$$

the trust radius is increased by some factor up to its maximum value $\overline{\Delta}$. The general method is given in Algorithm 19. In the algorithm the factors $1/2, 2$ for decreasing and increasing the trust radius are fixed. However, other values can be used.

The approximate function $m_k(x)$ can be minimized by various methods. As in the case of line search, we do not necessarily need the exact optimal solution. An easy method is to minimize the linear approximation, $\min_{\|r\| < \Delta}\{f(x_k) + \nabla f(x_k)^T r\}$. Its solution is the steepest descent direction,

---

Algorithm 19 TrustRegion($\overline{\Delta}, f, m_k, x_0, \mu, \nu$)

$k := 1, \Delta := \overline{\Delta}$
while (termination condition does not fulfill)
    $r_k := \text{argmin}_{\|r\|<\Delta} \, m_k(x_k + r)$
    if $(\nu < \rho_k)$
        $\Delta := \max\{2\Delta, \overline{\Delta}\}$
    else
            while $(\rho_k < \mu)$
                $\Delta := \Delta/2$
                $r_k := \text{argmin}_{\|r\|<\Delta} \, m_k(x_k + r)$
            endwhile
    $x_{k+1} := x_k + r_k$
    $k := k + 1$
endwhile

$r = -\nabla f(x_k)/\|\nabla f(x_k)\|$, where one only has to minimize the step length bounded to be less than the trust radius. The optimal step size can be given directly. Consider $r_k = \lambda r$, where $\|r\| = 1$ is normalized. When $r^T H_f(x_k)r \leq 0$, $m_k(x+\lambda r)$ is concave (or linear), descending in the direction of r. So the optimal step size is Δ. If it is convex, the minimum is taken either at the stationary point, where $\frac{\partial m_k(x+\lambda r)}{\partial \lambda} = \nabla f(x_k)^T r + \lambda r^T H_f(x_k)r = 0$, $(\lambda = \frac{-\nabla f(x_k)^T r}{r^T H_f(x_k)r})$, or at the maximum step size Δ, when the stationary point is outside;

$$\lambda = \begin{cases} \Delta & \text{if } r^T H_f(x_k)r \leq 0, \\ \min\left\{\dfrac{\|\nabla f(x_k)\|}{r^T H_f(x_k)r}, \Delta\right\}, & \text{otherwise.} \end{cases} \quad (5.20)$$

Notice that in this case the method is following a steepest descent method with a bounded line search. Consequently, the convergence near the optima is similar to that of the steepest descent method.

Example 5.13. Consider the problem in Example 5.9 with $f(x) = (x_1 - 3)^2 + 3(x_2 - 1)^2 + 2$ and $x_0 = (0,0)^T$. The initial trust radius is taken $\Delta_0 = 1$, and the maximum trust radius $\overline{\Delta} = 2$. The first direction is $(1,1)^T$ as in Example 5.9. Now the step length is $\lambda = 1$ according to (5.20). This gives as next iterate $x_1 = (\frac{1}{\sqrt{2}}, \frac{1}{\sqrt{2}})$. The function to minimize is quadratic, so the predicted reduction is the same as the actual one. For formula (5.19) this means that $\forall k \ \rho_k = 1$ and so $\Delta = 2$. In the rest of the steps the trust radius is always greater than the optimal step size. The iterates follow the steepest descent algorithm from this point. The run is depicted in Figure 5.14.

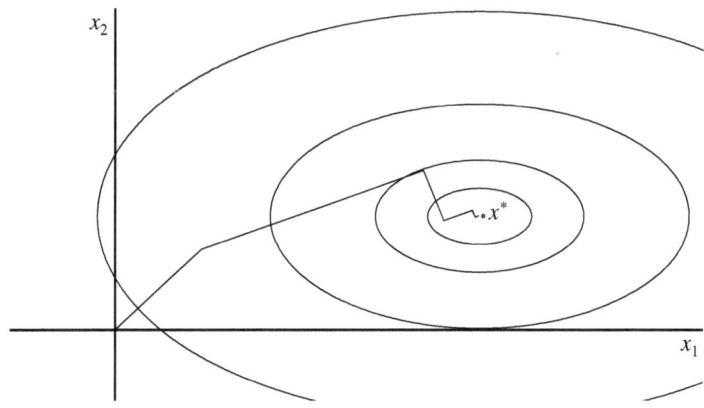

Fig. 5.14. Trust region method on the function $f(x) = (x_1 - 3)^2 + 3(x_2 - 1)^2 + 2$ with $x_0 = (0,0)^T$ and $\Delta_0 = 1$

Other approaches to solve (5.18) are the Dogleg method using Newton direction and the Steihaug approach with Levenberg–Marquardt idea. Also, the conjugate gradient method has a trust region version. For details, see, e.g., Kelley (1999) and Nocedal and Wright (2006).

5.5 Algorithms for nonlinear regression

The least squares problem of minimizing $f(\beta) = \sum_{i=1}^{m} (z(x_i, \beta) - y_i)^2$ as introduced in Section 2.7 has specific characteristics. Therefore, specific optimization methods have been developed to minimize $f(\beta)$. An important special case is that of linear regression, where z is linear in β. For ease of notation we will describe the linear regression case as $z(x, \beta) = x^T \beta$ and elaborate the minimization of its least squares around an example in Section 5.5.1.

The methods are based on the shape of the gradient and Hessean of $f(\beta)$. A useful concept is that of the so-called Jacobian being the $m \times n$ matrix of partial derivatives with elements

$$J_{ij}(\beta) = \frac{\partial z(x_i, \beta)}{\partial \beta_j}(\beta). \tag{5.21}$$

The partial derivatives of f are $\frac{\partial f(\beta)}{\partial \beta_j}(\beta) = 2 \sum_{i=1}^{m} \frac{\partial z(x_i, \beta)}{\partial \beta_j}(\beta)(z(x_i, \beta) - y_i)$. With the aid of the Jacobian and the error vector $e(\beta)$ with elements $e_i = z(x_i, \beta) - y_i$ they can be summarized as

$$\nabla f(\beta) = 2J^T(\beta)e(\beta). \tag{5.22}$$

The Hessean of f obtains a more sophisticated shape:

$$H_f(\beta) = 2J^T(\beta)J(\beta) + 2 \sum_{i=1}^{m} H_i(\beta)e_i(\beta), \tag{5.23}$$

where $H_i(\beta)$ is now the Hessean of the error $e_i(\beta)$ of the ith observation. The specific shape of gradient and Hessean gives rise to dedicated methods for optimizing f that are described in the following subsections.

5.5.1 Linear regression methods

The optimization of the sum of absolute values $f(\beta) = \sum |x_i^T \beta - y_i|$ or the infinite norm (maximum error) $f(\beta) = \max_i |x_i^T \beta - y_i|$ can be written as Linear Programming. The least squares criterion leads to the minimization of a quadratic function. Let us first of all remark that the Jacobian is a constant matrix X which does not depend on the parameter values in β. Thus we can write the least squares criterion in linear regression as

$$f(\beta) = (X\beta - y)^T(X\beta - y) = \beta^T X^T X\beta - 2y^T X\beta + y^T y, \tag{5.24}$$

which is a quadratic function in β. If the columns of X are linearly dependent, the minimum points can be found on a lower-dimensional plane. If they are independent, the minimum point is the stationary point of (5.24):

$$\beta^* = (X^T X)^{-1} X^T y \qquad (5.25)$$

if we follow equation (3.21). Notice that the same follows from finding a stationary point; $\nabla f(\beta) = 2X^T(X\beta - y) = 0$. The Hessean $2X^T X$ is positive semidefinite and its inverse has an important interpretation in statistics where the so-called variance–covariance matrix of the estimated β^* is proportional to $(X^T X)^{-1}$; see Bates and Watts (1988). The ellipsoidal level sets $f(\beta) - f(\beta^*) = (\beta - \beta^*)^T X^T X (\beta - \beta^*) < \delta$ have the interpretation of confidence regions in statistics.

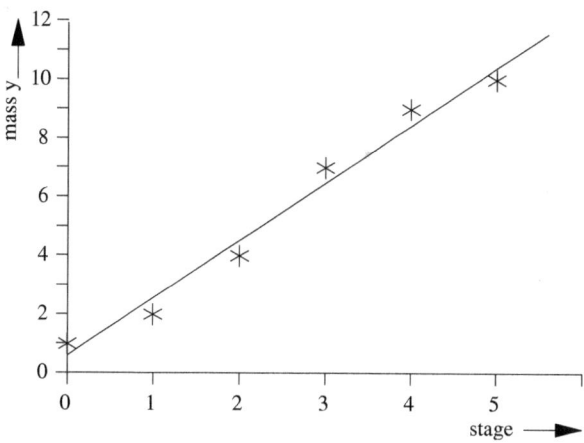

Fig. 5.15. Observations and estimated model of mass as function of stage

Example 5.14. We want to explain the mass y of a plant from its growth stage with a simple linear model $y = \beta_1 + \beta_2 stage$. The observed data points of stage 0 to 5 are given by $y = (1, 2, 4, 7, 9, 10)^T$. The X matrix is given by

$$X = \begin{pmatrix} 1\ 1\ 1\ 1\ 1\ 1 \\ 0\ 1\ 2\ 3\ 4\ 5 \end{pmatrix}^T,$$

such that

$$X^T X = \begin{pmatrix} 6 & 15 \\ 15 & 55 \end{pmatrix} \quad \text{and} \quad X^T y = \begin{pmatrix} 33 \\ 117 \end{pmatrix}.$$

Following (5.25) gives the least squares estimate

$$\beta^* = \begin{pmatrix} 6 & 15 \\ 15 & 55 \end{pmatrix}^{-1} \begin{pmatrix} 33 \\ 117 \end{pmatrix} = \begin{pmatrix} 0.57 \\ 1.97 \end{pmatrix}.$$

The corresponding model is $y = z(stage, \beta^*) = 0.57 + 1.97 stage$. Data points and model are illustrated in Figure 5.15.

5.5.2 Gauss–Newton and Levenberg–Marquardt

The method of Newton for least squares functions is given by

$$\beta_{k+1} = \beta_k - 2H_f^{-1}J^T(\beta_k)e(\beta_k), \tag{5.26}$$

where H_f is defined by the complicated expression (5.23). As it would be complicated to evaluate all Hesseans H_i in (5.23), one can use approximations with either the idea that z is linear in β or the idea that the error terms e_i are small.

The concept of the Gauss–Newton method is to approximate H_f by the first part $2J^T(\beta_k)J(\beta_k)$. Alternatively, one can say that the model z is linearized around β_k. The resulting search direction of Gauss–Newton is

$$r_k = -(J^T(\beta_k)J(\beta_k))^{-1}J^T(\beta_k)e(\beta_k), \tag{5.27}$$

which is a descent direction as $J^T J$ is a positive semidefinite matrix. It can be shown that for many instances, taking the final step sizes as 1 leads to convergence.

Example 5.15. A researcher investigates the effect of dosing two nutrients on the yield of tomatoes. Therefore he performs four experiments in separated fields. The resulting data are given in Table 5.8. The expected relation is

$$yield = (1 + \beta_1 dose_1)(1 + \beta_2 dose_2), \tag{5.28}$$

where β_1 and β_2 are reaction parameters. The least squares function to be optimized is $f(\beta) = \sum_1^4((1 + \beta_1 dose_{1i})(1 + \beta_2 dose_{2i}) - yield_i)^2$. The resulting Jacobian has rows $(dose_{1i}(1 + \beta_2 dose_{2i}), dose_{2i}(1 + \beta_1 dose_{1i}))$. Consider

Table 5.8. Observed yield of tomatoes and nutrient dosage

experiment	1	2	3	4
dose 1	1.0	1.0	1.0	2.0
dose 2	0.0	1.0	2.0	0.0
yield	0.5	5.0	6.5	1.0

starting vector $\beta_0 = (1,1)^T$, with sum of squared errors $f(\beta_0) = 19.5$. The error vector itself is $e(\beta_0) = (1.5, -1, -0.5, 2)^T$ and the Jacobian

$$J_0 = J(\beta_0) = \begin{pmatrix} 1 & 2 & 3 & 2 \\ 0 & 2 & 4 & 0 \end{pmatrix}^T \quad \text{and} \quad J_0^T J_0 = \begin{pmatrix} 18 & 16 \\ 16 & 20 \end{pmatrix},$$

such that the resulting steepest descent direction is

$$r = -\nabla f(\beta_0) = -2J^T(\beta_0)e(\beta_0) = \begin{pmatrix} -4 \\ 8 \end{pmatrix}.$$

The Gauss–Newton direction is determined by

$$r = - \begin{pmatrix} 18 & 16 \\ 16 & 20 \end{pmatrix}^{-1} \begin{pmatrix} 2 \\ -4 \end{pmatrix} = \begin{pmatrix} -1 \\ 1 \end{pmatrix}.$$

This is a descent direction, as it makes a sharp angle with $-\nabla f(\beta_0)$.

One of the most used algorithms is due to Levenberg–Marquardt, which has been implemented in most statistical software (Marquardt, 1963). The basic iteration scheme is based on

$$\beta_{k+1} = \beta_k - (J^T(\beta_k)J(\beta_k) + \alpha_k E)^{-1}J^T(\beta_k)e(\beta_k), \qquad (5.29)$$

where E is the unit matrix and α_k implicitly determines the step size. For α big, the method follows the steepest descent. For smaller α, it looks more like the Gauss–Newton method. Usually a scheme is followed where the size of α_k is reduced during the iterations.

5.6 Algorithms for constrained optimization

We write the generic NLP problem now as

$$\begin{aligned} &\min f(x) \\ \text{s.t.} \quad &g_i(x) \le 0 \; i = 1, \ldots, p, \text{ inequality constraints,} \\ &g_i(x) = 0 \; i = p+1, \ldots, m, \text{ equality constraints.} \end{aligned} \qquad (5.30)$$

Until now we have ignored the presence of the constraints g and searched for the optimum in the whole space. When dealing with constraints, there are two main options to take. One is to convert the problem into unconstrained problem(s) by embedding the constraints in the objective function, or directly restricting the search to the feasible area. In the first case, the new unconstrained problems are not equivalent to the original problem, but using some parameters, their solutions tend to the solution of the constrained problem. In this way the previously discussed methods can be used to solve these new problems. In this type of method, the question is how to embed the constraints in the objective. We will discuss the penalty and barrier function method in Section 5.6.1.

In the other case, directly restricting the search to the feasible area, we usually modify an unconstrained method. Starting from a feasible point, the direction and step length of the original method are modified such that the new point is also feasible. Such methods are the gradient projection method and sequential quadratic programming discussed in Sections 5.6.2 and 5.6.3.

5.6.1 Penalty and barrier function methods

The penalty function method was introduced by Zangwill (1967) and also by Pietrzykowski (1969). The main idea of the method is to penalize infeasibility. The penalty functions

$$p_\mu(x) = \mu \left(\sum_{i=1}^{p} \max\{g_i(x), 0\} + \sum_{i=p+1}^{m} |g_i(x)| \right)$$

and

$$p_\mu(x) = \mu \left(\sum_{i=1}^{p} (\max\{g_i(x), 0\})^2 + \sum_{i=p+1}^{m} g_i^2(x) \right)$$

are 0 when x is feasible, but take a positive value at infeasible points. Adding the penalty function to the objective function, $P_\mu(x) = f(x) + p_\mu(x)$, we get an unconstrained problem for every value of μ,

$$\min P_\mu(x). \tag{5.31}$$

It means that the objective function of the converted unconstrained problem has high values at infeasible areas. The minimizer of (5.31) approximates the minimizer of (5.30) for a value of μ that is high enough. However, it is not known apriori how high μ should be. The minimizer can be far from feasibility even for a relatively high μ value. Moreover, choosing a high value for μ can result in a so-called ill-conditioned problem. It means that the penalty function has values much larger in order of magnitude than $f(x)$. Numerical methods can fail or give false results in such cases.

To resolve this problem, the penalty function method works as follows (see Algorithm 20). Solve the penalized unconstrained problem $\min P_\mu(x)$ for a given value for μ. If the minimizer $x^*(\mu)$ fulfills $p_\mu(x^*(\mu)) \leq \epsilon$, $x^*(\mu)$ is accepted as an approximate solution. Otherwise the value of μ is increased and the penalized unconstrained problem solved until the above condition is fulfilled. The minimization of the next unconstrained problem starts from the last minimum, to reach the solution in fewer steps. Moreover, one prevents ill-conditioning in the neighborhood of the optimization path.

Algorithm 20 PenaltyMethod($f, g, p, \mu_0, \beta, \epsilon$)

$k := 0$
$x_k := \operatorname{argmin} P_\mu(x)$
while $(p_\mu(x_k) > \epsilon)$
$\quad k := k + 1$
$\quad \mu_k := \beta \cdot \mu_{k-1}$
$\quad x_k := \operatorname{argmin} P_\mu(x)$
endwhile

Example 5.16. Consider the problem

$$\min \ 5 - e^x$$

$$\text{s.t.} \ \ x = 1.$$

The constraint defines minimum point $x^* = 1$. Taking $p_\mu(x) = \mu(x-1)^2$, the unconstrained problem is

$$\min\{5 - e^x + \mu(x-1)^2\}.$$

Setting $\mu_0 = 1$ and $\beta = 2$, the objective function of the first four unconstrained problems is depicted in Figure 5.16. The solution x_k tends to 1 as μ_k goes to infinity.

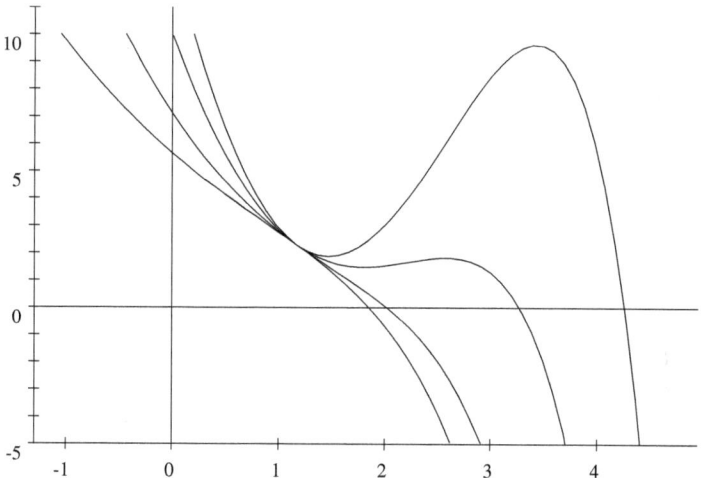

Fig. 5.16. The functions $P_\mu(x)$ for $\mu = 1, 2, 4, 8$

Example 5.17. The penalty function method is used to find the solution of

$$\min\ x_1^2 + x_2^2$$
$$\text{s.t.}\ \ x_1 + x_2 = 2.$$

Using the quadratic penalty function, we minimize

$$P_\mu(x_1, x_2) = x_1^2 + x_2^2 + \mu(x_1 + x_2 - 2)^2.$$

The first-order necessary conditions in minimum point $x^*(\mu)$ are

$$\frac{\partial P_\mu}{\partial x_1} = 0 \qquad \frac{\partial P_\mu}{\partial x_2} = 0.$$

Thus, $2x_1 + \mu 2(x_1 + x_2 - 2) = 0$ and $2x_2 + \mu 2(x_1 + x_2 - 2) = 0$, from which $x_1 = x_2 = \frac{2\mu}{2\mu+1}$. In Table 5.9 we can see that x_k tends to the solution $x^* = (1, 1)$ if the unconstrained problems are exactly solved using $\mu_0 = 1$ and $\beta = 4$.

Table 5.9. Steps by the penalty function method for Example 5.17

k	μ	x_k
0	1	(0.7500, 0.7500)
1	4	(0.8888, 0.8888)
2	16	(0.9696, 0.9696)
3	64	(0.9922, 0.9922)
4	256	(0.9980, 0.9980)
5	1024	(0.9995, 0.9995)
6	4096	(0.9998, 0.9998)

One can observe that the solution reached by the penalty function method and all subsequent points are infeasible. Therefore, in applications where feasibility is strictly required, penalty function methods cannot be used. In such cases barrier function methods are more appropriate.

Barrier functions make a barrier at the constraints such that x_k can only be situated in the interior of the feasible area. If the minimizer of the original problem is on the boundary of the feasible region, x_k tends to the boundary from the interior. It also means that the barrier function method works only with inequality constraints (there is no interior for an equality constraint). For instance, the barrier functions

$$b_\mu(x) = -\mu \sum_{i=1}^{p} \frac{1}{g_i(x)}$$

and

$$b_\mu(x) = -\mu \sum_{i=1}^{p} \ln(-g_i(x))$$

give positive values for strictly feasible points and infinity when $g_i(x) = 0$ for some i. Note that the barrier function at infeasible points is not necessarily defined. In contrast to the penalty function method we do have to take care not to leave the feasible area while minimizing $B_\mu(x) = f(x) + b_\mu(x)$. One could think that in this way the problem did not become easier as we still have the constraints to be taken into account. Although the latter is true,

---

Algorithm 21 BarrierMethod($f, g, b, \mu_0, \beta, \epsilon$)

$k := 0$
$x_k := \text{argmin}_{x \in X} B_\mu(x)$
while ($b_\mu(x_k) > \epsilon$)
    $k := k + 1$
    $\mu_k := \frac{\mu_{k-1}}{\beta}$
    $x_k := \text{argmin}_{x \in X} B_\mu(x)$
endwhile

---

for the new problems none of the constraints are active, so any unconstrained method can be used with some safeguards.

In Algorithm 21 a general barrier function method is given. The algorithm is mainly the same as the penalty function method except that here μ tends to zero in order to have $B_\mu(x) \to f(x)$.

Example 5.18. Consider the barrier function method for a variant of the problem in Example 5.17,

$$\min \ x_1^2 + x_2^2$$
$$\text{s.t. } x_1 + x_2 \geq 2.$$

Using the logarithmic barrier function, our new problem is to minimize $B_\mu(x_1, x_2) = x_1^2 + x_2^2 - \mu \ln(x_1 + x_2 - 2)$. The solution must satisfy the first-order optimality condition, that is,

$$\frac{\partial B_\mu}{\partial x_1} = 2x_1 - \mu \frac{1}{x_1 + x_2 - 2} = 0 \qquad \frac{\partial B_\mu}{\partial x_2} = 2x_2 - \mu \frac{1}{x_1 + x_2 - 2} = 0.$$

Solving these equations, we get that $x^*(\mu) = (\frac{1}{2} + \frac{1}{2}\sqrt{1+\mu}, \frac{1}{2} + \frac{1}{2}\sqrt{1+\mu})$. In Table 5.10 the run of the barrier function method is given for $\mu_0 = 1$ and $\beta = 2$. We assume the exact optimum is found by the local optimizer.

Table 5.10. Steps by the barrier function method for Example 5.18

k	μ	x_k
0	1	(1.2071, 1.2071)
1	0.5	(1.1123, 1.1123)
2	0.25	(1.0590, 1.0590)
3	0.125	(1.0303, 1.0303)
4	0.0625	(1.0153, 1.0153)
5	0.03125	(1.0077, 1.0077)
6	0.015625	(1.0038, 1.0038)

For the barrier function method every subproblem is ill-conditioned, as B_μ is unbounded at the constraints. Hence, the logarithmic barrier function is used generally, as it grows in a less dramatic way than $\frac{1}{x}$. Because of the ill-conditionedness, the above methods are not prevalent. In the following we will discuss more practical methods.

5.6.2 Gradient projection method

This method is a modification of the steepest descent method (see Section 5.4.1) for constrained optimization. It was developed in the early 1960s by

Rosen (1960, 1961) and later improved by Haug and Arora (1979). At every step the new direction is modified in order to stay in the feasible region by projecting the gradient to the active constraints. In Figure 5.17 the negative gradient of the objective $-\nabla f(x)$, the constraint $g(x)$ and its gradient $\nabla g(x)$ are depicted together with the projected direction r.

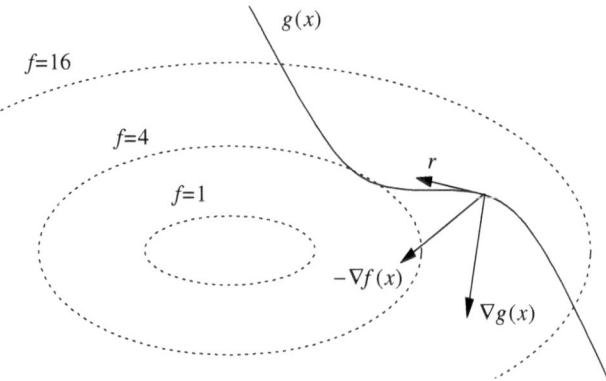

Fig. 5.17. The projected gradient direction

The projection is done by a projection matrix, that is, $r = -P\nabla f$. Let M be the Jacobian matrix of the active constraints; it consists of column vectors $\nabla g_i(x)$ for these constraints for which $g_i(x) = 0$. The projection matrix can be computed as

$$P = E - M(M^T M)^{-1} M^T.$$

How do we get this formula? We know that for every active constraint the direction r is perpendicular to its gradient, $\nabla g_i^T r = 0$, such that

$$M^T r = 0.$$

The steepest descent direction along the binding constraint can be obtained by solving the problem

$$
\begin{aligned}
\min \ & r^T \nabla f \\
\text{s.t.} \ & M^T r = 0, \\
& ||r||_2 = 1.
\end{aligned}
\qquad (5.32)
$$

That is, we are searching for the most negative direction, which has unit length. Using the Lagrangean (see Section 3.5.1) of (5.32),

$$L(r, u, v) = r^T \nabla f + r^T M u + v(r^T r - 1),$$

where $u \in \mathbb{R}^n, v \in \mathbb{R}, ||r||_2 = r^T r$, the necessary condition for optimality is

$$\frac{\partial L}{\partial r} = \nabla f + Mu + 2vr = 0. \tag{5.33}$$

Multiplying (5.33) by M^T and considering $M^T r = 0$,

$$M^T \nabla f + M^T M u + 2v M^T r = M^T \nabla f + M^T M u = 0,$$

from which

$$u = -(M^T M)^{-1} M^T \nabla f.$$

Substituting in (5.33) gives the projected direction

$$r = -\frac{1}{2v}(E - M(M^T M)^{-1} M^T)\nabla f.$$

The factor $\frac{1}{2v}$ can be omitted, as r stands for a direction. Recall that the step length is determined by the line search. When $r = 0$ and $u \geq 0$ the Kuhn–Tucker conditions are satisfied, thus we have found a KKT point. If some Lagrange multipliers are negative ($u_i < 0$ for some i), that means we may still find a decreasing direction by removing constraints with $u_i < 0$. In fact the negative multiplier means that the corresponding constraint is not binding for the decreasing direction. Usually, first the constraint with the most negative Lagrange multiplier is removed from the active constraints and r is calculated again. If $r \neq 0$, a decreasing direction is found. Otherwise we remove more constraints with negative Lagrange multipliers. If there is no more $u_i < 0$, but $r = 0$, we can stop. We have reached a point where the Karush–Kuhn–Tucker conditions hold.

After finding a feasible direction r, we want to obtain the optimal step length $\lambda = \operatorname{argmin}_{\mu>0} f(x_k + \mu r)$, such that the new iterate fulfills the nonbinding constraints, i.e., $g_i(x_k + \lambda r) \leq 0$. In fact, the constraint that becomes binding first along direction r determines the maximum step length $\lambda_{\max}$. Specifically for a linear constraint $a_i^T x - b_i \leq 0$, λ should satisfy $a_i^T(x_k + \lambda r) - b_i \leq 0$, such that $\lambda_{\max} \leq \frac{b_i - a_i^T x_k}{a_i^T r}$ over all linear constraints. The main procedure is elaborated in Algorithm 22 for the case where only linear constraints exist.

Example 5.19. Consider the problem

$$
\begin{aligned}
\min \quad & x_1^2 + x_2^2 \\
\text{s.t.} \quad & x_1 + x_2 \geq 2, \\
& -2x_1 + x_2 \leq 1, \\
& x_1 \geq \tfrac{1}{2}.
\end{aligned}
$$

Let x_0 be $(0.5, 2)^T$. The gradient is $\nabla f(x) = (2x_1, 2x_2)^T$, so at x_0 we have $\nabla f(x_0) = (1, 4)^T$. We can see that the second and third constraint are active, but not the first. Thus, $M = \begin{pmatrix} -2 & -1 \\ 1 & 0 \end{pmatrix}$, $(M^T M)^{-1} = \begin{pmatrix} 1 & -2 \\ -2 & 5 \end{pmatrix}$, and we

---

Algorithm 22 GradProj(f, g, x_0, ϵ)

$k := 0$

do

    $r := -(E - M(M^T M)^{-1} M^T) \nabla f$

    while ($r = 0$)

        $u := -(M^T M)^{-1} M^T \nabla f$

        if ($\min_i u_i < 0$)

            Remove g_i from the active constraints and recalculate r

        else

            return x_k (a KKT point)

    endwhile

    $\lambda := \text{argmin}_\mu \, f(x_k + \mu r)$

    if $\exists i \; g_i(x_k + \lambda r) < 0$

        Determine $\lambda_{\max}$

        $\lambda = \lambda_{\max}$

    $x_{k+1} := x_k + \lambda r$

    $k := k + 1$

while($|x_k - x_{k-1}| > \epsilon$)

---

get $P = \begin{pmatrix} 0 & 0 \\ 0 & 0 \end{pmatrix}$. Hence, $r = 0$. Now, computing the Lagrangean coefficients $u = (-4, 9)$, we can see that the second constraint (with coefficient -4) does not bind the steepest descent direction, so that should not be considered in the projection. Thus, $M = \begin{pmatrix} -1 \\ 0 \end{pmatrix}$, $P = \begin{pmatrix} 0 & 0 \\ 0 & 1 \end{pmatrix}$ and $r = \begin{pmatrix} 0 \\ -4 \end{pmatrix}$. We can normalize to $r = (0, -1)^T$ and compute the optimal step length λ. One can check that the minimum of $f(x_k + \lambda r)$ is 2, but the originally nonbinding constraint, g_1, is not fulfilled with such a step. To satisfy $g_1(x_k + \lambda r) \geq 0$, the maximum step length 0.5 is taken, so $x_1 = (0.5, 1.5)^T$.

Now the two binding constraints are g_1 and g_3, while $\nabla f(x_1) = (1, 3)^T$. Corresponding $M = \begin{pmatrix} -1 & -1 \\ -1 & 0 \end{pmatrix}$ is nonsingular, $P = 0$ and $r = 0$. Checking the Lagrangeans we get $u = (3, -2)^T$, which means g_3 does not have to be considered in the projection. With the new $M = (-1, -1)^T$ the projection matrix $P = \frac{1}{2} \begin{pmatrix} 1 & -1 \\ -1 & 1 \end{pmatrix}$, and so $r = (1, -1)^T$. The optimal step length $\lambda = \text{argmin}_\mu \, f(x_k + \mu r) = 0.5$, with which $x_2 = (1, 1)^T$ satisfies all the constraints. One can check that x_2 is the optimizer (a KKT point) by having $P = 0$ and $u \geq 0$. The problem and the steps are depicted in Figure 5.18.

For nonlinear constraints an estimate of the maximum value of λ can be calculated using the linear approximations of the constraints. Another approach is to use a desired reduction of the objective, like $f(x_k) - f(x_{k+1}) \approx \gamma \cdot f(x_k)$. Using this assumption we get directly the step length; see Haug and Arora (1979).

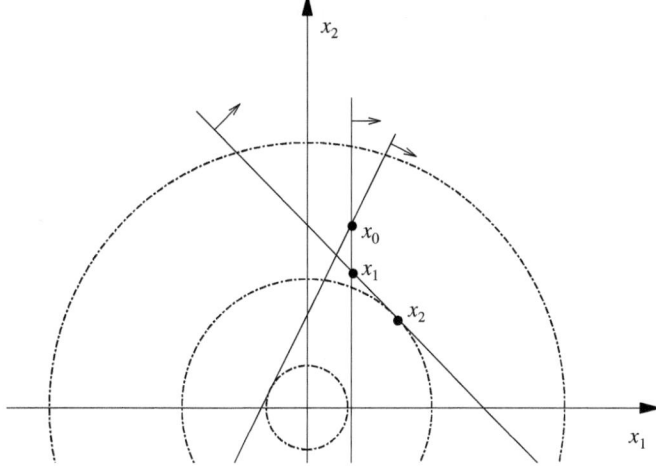

Fig. 5.18. The steps for Example 5.19

In case of nonlinear constraints, we also have to take care that the new iterate is not violating the active constraints. As we are moving perpendicular to the gradients of the constraints, we may need to do a restoration move to get back to the feasible area as illustrated in Figure 5.19.

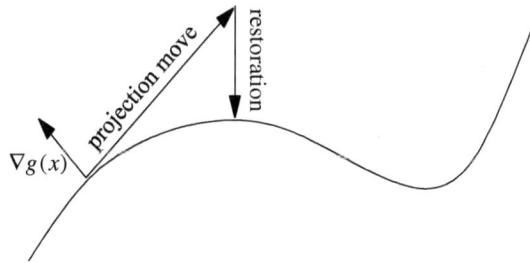

Fig. 5.19. The projected and the restoration move

The idea of projecting the steepest descent can be generalized for other descent direction methods. One simply has to change $-\nabla f$ to the desired direction in Algorithm 22 to obtain the projected version of a descent direction method.

In the next section we are going to discuss the sequential quadratic programming which is also called the projected Lagrangean method.

5.6.3 Sequential quadratic programming

To our knowledge SQP was first introduced in the Ph.D. thesis of Wilson (1963), later modified by Han (1976) and Powell (1978). SQP can be viewed as a modified Newton method for constrained optimization. Actually it is a Newton method applied to the KKT conditions. Using the method, a sequence of quadratic programming problem is solved. That is, at every iteration the quadratic approximation of the problem is solved, namely, the quadratic approximation of the Lagrangean function with the linear approximation of the constraints.

Let us start with equality constrained problems,

$$
\begin{aligned}
\min \quad & f(x) \\
\text{s.t.} \quad & g(x) = 0.
\end{aligned}
\tag{5.34}
$$

The KKT conditions for (5.34) are

$$
\begin{aligned}
\nabla f(x) + u\nabla g(x) &= 0 \\
g(x) &= 0.
\end{aligned}
\tag{5.35}
$$

Observe that the first KKT equation says the gradient (with respect to the x-variables) of the Lagrangean should be zero, i.e., $\nabla_x L(x, u) = 0$. In Section 5.4.2, we discussed that the Newton method can be used to determine a stationary point. To work with the same idea, we define $\nabla_x^2 L(x, u)$ as the Hessean of the Lagrangean with respect to the x-variables. To solve (5.35), the iterates are given by $x_{k+1} = x_k + r, u_{k+1} = u_k + v$, where r, v are the solutions of

$$
\begin{pmatrix} \nabla_x^2 L(x_k, u_k) & \nabla g(x_k) \\ \nabla g(x_k)^T & 0 \end{pmatrix} \begin{pmatrix} r \\ v \end{pmatrix} = - \begin{pmatrix} \nabla_x L(x_k, u_k) \\ g(x_k)^T \end{pmatrix}.
\tag{5.36}
$$

Example 5.20. Consider the problem

$$
\begin{aligned}
\min \quad & (x_1 - 1)^2 + (x_2 - 3)^2 \\
\text{s.t.} \quad & x_1 = x_2^2 - 1.
\end{aligned}
$$

Our constraint is $g(x) = -x_1 + x_2^2 - 1 = 0$ and the Lagrangean is $L(x, u) = (x_1 - 1)^2 + (x_2 - 3)^2 + u(x_1 - x_2^2 + 1)$. The gradients are $\nabla_x L(x, u) = (2(x_1 - 1) + u, 2(x_2 - 3) - 2x_2 u)^T$ and $\nabla g(x) = (-1, 2x_2)^T$, and the Hessean for L is $\nabla_x^2 L(x, u) = \begin{pmatrix} 2 & 0 \\ 0 & 2 - 2u \end{pmatrix}$.

Denoting by N the matrix of (5.36) and by rhs the right-hand-side vector, we have $N = \begin{pmatrix} 2 & 0 & -1 \\ 0 & 2 - 2u & 2x_2 \\ -1 & 2x_2 & 0 \end{pmatrix}$ and $rhs = \begin{pmatrix} 2(1 - x_1) - u \\ 2(3 - x_2) + 2x_2 u \\ x_1 - x_2^2 + 1 \end{pmatrix}$.

Consider as starting point $x_0 = (0,0)^T$ and starting value for the multiplier $u_0 = 2$. This gives $N_0 = \begin{pmatrix} 2 & 0 & -1 \\ 0 & 6 & 0 \\ -1 & 0 & 0 \end{pmatrix}$ and $rhs_0 = \begin{pmatrix} 4 \\ 6 \\ -1 \end{pmatrix}$ giving a solution of (5.36) of $(r^T, v) = (1,1,-2)$, such that $x_1 = (1,1)^T$ and $u_1 = 0$. Following this process, $N_1 = \begin{pmatrix} 2 & 0 & -1 \\ 0 & 2 & 2 \\ -1 & 2 & 0 \end{pmatrix}$ and $rhs_1 = \begin{pmatrix} 0 \\ 4 \\ -1 \end{pmatrix}$.

Now $(r^T, v) = (1,0,2)$, such that we reach the optimum point $x_2 = (2,1)^T$ with $u_2 = 2$. This point fulfills the KKT conditions.

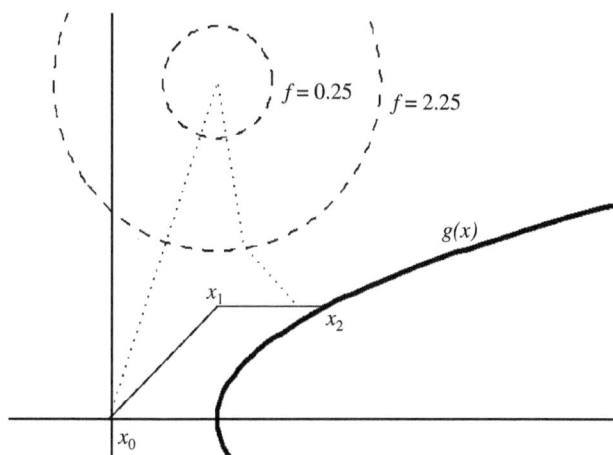

Fig. 5.20. Iterates in Example 5.20

Figure 5.20 shows the constraint, contours and the iterates. Moreover, a second process is depicted which starts from the same starting point $x_0 = (0,0)^T$, but takes for the multiplier $u_0 = 0$. One can verify that more iterations are needed.

Applying the same idea to inequality constrained problems requires more refinement; one has to take care of complementarity and the nonnegative sign of the multipliers.

5.7 Summary and discussion points

- Nonlinear programming methods can use different information on the instance to be solved; the fact that the function value is higher in different points, the value of the function, the derivative or second derivative.

- Interval methods based on bracketing, bisection and the golden section rule lead to a linear convergence speed.
- Interpolation methods like quadratic and cubic interpolation and the method of Newton are usually faster, require information of increased order and safeguards to force convergence for all possible instances.
- The method of Nelder–Mead and the Powell method can be used when no derivative information is available and even when functions are not differentiable. The latter method is usually more efficient, but we found the first more in implementations.
- Many NLP methods use search directions and one-dimensional algorithms to do line search to determine the step size.
- When (numerical) derivative information is used, the search direction can be based on the steepest descent, conjugate gradient methods and quasi-Newton methods.
- Nonlinear regression has specific methods that exploit the structure of the problem, namely, Gauss–Newton and Levenberg–Marquardt method.
- For constrained problems there are several approaches; using penalty approaches or dealing with the constraints in the generation of search directions and step sizes. In the latter the iterative identification of active (binding) constraints is a major task.

5.8 Exercises

1. Given $f(x) = (x^2 - 4)^2$, starting point $x_0 = 0$ and accuracy $\epsilon = 0.1$.
   (a) Generate with the bracketing algorithm an interval $[a, b]$ which contains a minimum point of f.
   (b) Apply the golden section algorithm to reduce $[a, b]$ to an interval smaller in size than ϵ which contains a minimum point.

2. Given Algorithm 23, function $f(x) = x^2 - 1.2x + 4$ on interval $[0, 4]$ and accuracy $\epsilon = 10^{-3}$.

Algorithm 23 Grid3($[a, b], f, \epsilon$)

Set $k := 1$, $a_1 := a$ and $b_1 := b$
$x_0 := (a + b)/2$, evaluate $f(x_0)$
while $(b_k - a_k > \epsilon)$
$\quad l := a_k + \frac{1}{4}(b_k - a_k)$, $r := a_k + \frac{3}{4}(b_k - a_k)$
$\quad$ evaluate $f(l)$ and $f(r)$
$\quad x_k := \operatorname{argmin}\{f(l), f(x_{k-1}), f(r)\}$
$\quad a_{k+1} := x_k - \frac{1}{4}(b_k - a_k)$, $b_{k+1} := x_k + \frac{1}{4}(b_k - a_k)$
$\quad k := k + 1$
endwhile

(a) Perform three iterations of the algorithm.

(b) How many iterations are required to reach the final accuracy?

(c) How many function evaluations does this imply?

3. Given Algorithm 24 for finding a minimum point of 2D function $f : \mathbb{R}^2 \rightarrow \mathbb{R}$, function $f(x) = 2x_1^2 + x_2^2 + 2\sin(x_1 + x_2)$ on interval $[a, b]$ with $a = (-1, -1)^T$ and $b = (1, 0)^T$ and accuracy $\epsilon = 10^{-3}$.

Algorithm 24 2DBisect($[a, b], f, \epsilon$)

Set $k := 0$, $a_0 := a$ and $b_0 := b$
while ($\|b_k - a_k\| > \epsilon$)
   $x_k := \frac{1}{2}(a_k + b_k)$
   Determine $\nabla f(x_k)$
   if $\frac{\partial f}{\partial x_1}(x_k) < 0$, $a_{k+1,1} := x_{k,1}$ and $b_{k+1,1} := b_{k,1}$
   else $a_{k+1,1} := a_{k,1}$ and $b_{k+1,1} := x_{k,1}$
   if $\frac{\partial f}{\partial x_2}(x_k) < 0$, $a_{k+1,2} := x_{k,2}$ and $b_{k+1,2} := b_{k,2}$
   else $a_{k+1,2} := a_{k,2}$ and $b_{k+1,2} := x_{k,2}$
   $k := k + 1$
endwhile

(a) Perform three iterations of the algorithm. Draw the corresponding intervals $[a_k, b_k]$ which enclose the minimum point.

(b) Give an estimate of the minimum point.

(c) How many iterations are required to reach the final accuracy?

4. Given function $f(x) = x_1^2 + 4x_1x_2 + x_2^2 + e^{x_1^2}$ and starting point $x_0 = (0, 1)^T$.

(a) Determine the steepest descent direction in x_0.

(b) Determine the Newton direction in x_0. Is this a descent direction?

(c) Is $H_f(x_0)$ positive definite?

(d) Determine the stationary points of f.

5. Given an NLP algorithm where the search directions are generated as follows, $r_0 := -\nabla f(x_0)$, the steepest descent and further $r_k := -M_k \nabla f(x_k)$, with $M_k := I + r_{k-1} r_{k-1}^T$, where I is the unit matrix.

(a) Show that M_k is positive definite.

(b) Show that r_k coincides with the steepest descent direction if exact line minimization is used to determine the step size.

6. Given quadratic function $f(x) = x_1^2 - 2x_1x_2 + 2x_2^2 + -2x_2$ and starting point $x_0 = (0, 0)^T$.

(a) Determine the steepest descent direction r_0 in x_0.

(b) Determine the step size in direction r_0 by line minimization.

(c) Given that M_0 is the unit matrix, determine M_1 via the BFGS update.

(d) Determine corresponding BFGS direction $r_1 = -M_1 \nabla f(x_1)$ and perform a line search in that direction.

(e) Show in general that the quasi-Newton condition holds for BFGS, i.e.,
$r_k = M_{k+1}y_k$.

7. Three observations are given, $x = (0, 3, 1)^T$ and $y = (1, 16, 4)^T$. One assumes the relation between x and y to be

$$y = z(x, \beta) = \beta_1 e^{\beta_2 x}. \tag{5.37}$$

(a) Give an estimate of β as minimization of the sum of $(y_i - z(x_i, \beta))^2$.
(b) Draw observations x_i, y_i and prediction $z(x_i, \beta)$ for $\beta = (1, 1)^T$.
(c) Determine the Jacobian $J(\beta)$.
(d) Determine the steepest descent direction in $\beta_0 = (1, 0)^T$.

8. Using the infinite norm in nonlinear regression leads to a nondifferentiable problem minimizing $f(\beta) = \max_i |y_i - z(x_i, \beta)|$. Algorithm 25 has been designed to generate an estimation of β given data $x_i, y_i, i = 1, \ldots, m$. In the algorithm, $J_i(\beta)$ is row i of the Jacobian. Data on the length x

---

Algorithm 25 Infregres($z, x, y, \beta_0, \epsilon$)

$k := 0$
repeat
    Determine $f(\beta_k) = \max_i |y_i - z(x_i, \beta_k)|$
    direction $r := 0$
    for $(i = 1, \ldots, m)$ **do**
        if $(y_i - z(x_i, \beta_k) = f(\beta_k))$
            $r := r + J_i(\beta)$
        if $(z(x_i, \beta_k - y_i) = f(\beta_k))$
            $r := r - J_i(\beta)$
    $\lambda := 5$
    while $(f(\beta_k + \lambda r_k) > f(\beta_k))$
        $\lambda := \frac{\lambda}{2}$
    endwhile
    $\beta_{k+1} := \beta_k + \lambda r_k$
    $k := k + 1$
until $(\|\beta_k - \beta_{k-1}\| > \epsilon)$

---

and weight y of four students is given; $x = (1.80, 1.70, 1.60, 1.75)^T$ and $y = (90, 80, 60, 70)^T$. The model to be estimated is $y = z(x, \beta) = \beta_1 + \beta_2 x$ and initial parameter values $\beta_0 = (0, 50)^T$.
(a) Give an interpretation of the while-loop in Algorithm 25. Give an alternative scheme for this loop.
(b) Draw in an x, y-graph the observations and the line $y = z(x, \beta_0)$.
(c) Give values β for which $f(\beta)$ is not differentiable.
(d) Perform two iterations with Algorithm 25 and start vector β_0. Draw the obtained regression lines $z(x, \beta_k)$ in the graph made for point (b).

(e) Give the formulation of an LP problem which solves the specific estimation problem of $\min_\beta f(\beta)$.

9. In order to find a feasible solution of a set of inequalities $g_i(x) \le 0$, $i = 1, \ldots, m$, one can use a penalty approach in minimizing $f(x) = \max_i g_i(x)$.
   (a) Show with the definition that f is convex if g_i is convex for all i.
   (b) Given $g_1(x) = x_1^2 - x_2$, $g_2(x) = x_1 - x_2 + 2$. Draw the corresponding feasible area in $\mathbb{R}^2$.
   (c) Give a point x for which $f(x)$ is not differentiable.
   (d) For the given set of inequalities, perform two iterations with Algorithm 26 and start vector $x_0 = (1, 0)$.
   (e) Do you think Algorithm 26 always converges to a solution of the set of inequalities if a feasible solution exists?

Algorithm 26 feas$(x_0, g_i(x)$, $i = 1, \ldots, m)$

Set $k := 0$, determine $f(x_0) = \max_i g_i(x_0)$
while $(f(x_k) > 0)$
    determine an index $j \in \text{argmax}_i\, g_i(x_k)$
    search direction $r_k := -\nabla g_j(x_k)$
    $\lambda := 1$
    while $(f(x_k + \lambda r_k) > f(x_k))$
        $\lambda := \frac{\lambda}{2}$
    endwhile
    $x_{k+1} := x_k + \lambda r_k$
    $k := k + 1$
endwhile

---

10. Linear Programming is a special case of NLP. Given problem

$$\max_X f(x) = x_1 + x_2, \quad X = \{x \in \mathbb{R}^2 | 0 \le x_1 \le 4, 0 \le x_2 \le 3\}. \quad (5.38)$$

An NLP approach to solve LP is to maximize a so-called logbarrier function $B_\mu(x)$ where one studies $\mu \to 0$. In our case

$$B_\mu(x) = x_1 + x_2 + \mu(\ln(x_1) + \ln(x_2) + \ln(4 - x_1) + \ln(3 - x_2)). \quad (5.39)$$

Given points $x_0 = (4, 1)^T$ and $x_1 = (1, 1)^T$.
   (a) Show that x_0 does not fulfill the KKT conditions of problem (5.38).
   (b) Give a feasible ascent direction r in x_0.
   (c) Is $f(x)$ convex in direction r?
   (d) For which values of $x \in \mathbb{R}^2$ is B_μ defined?
   (e) $\mu = 1$, determine the steepest ascent direction in x_1.
   (f) $\mu = 1$, determine the Newton direction in x_1.
   (g) Determine the stationary point $x^*(\mu)$ of B_μ.
   (h) Show that the KKT conditions are fulfilled by $\lim_{\mu \to 0} x^*(\mu)$.

(i) Show that B_μ is concave on its domain.

11. Given optimization problem $\max_X f(x) = (x_1 - 1)^2 + (x_2 - 1)^2$, $X = \{x \in \mathbb{R}^2 | 0 \le x_1 \le 6, 0 \le x_2 \le 4\}$ and $x_0 = (3,2)^T$. One can try to obtain solutions by maximizing the so-called shifted logbarrier function $G_\mu(x) = f(x) + \mu \sum_i \ln(-g_i(x) + 1)$, which in this case is

$$G_\mu(x) = (x_1-1)^2 + (x_2-1)^2 + \mu(\ln(x_1+1) + \ln(x_2+1) + \ln(7-x_1) + \ln(5-x_2)).$$

(a) For which values of $x \in \mathbb{R}^2$ is G_μ defined?
(b) Determine the steepest ascent direction of $G_3(x)$ in x_0.
(c) Determine the Newton direction of $G_3(x)$ in x_0.
(d) For which values of μ is G_μ concave around x_0?

12. Find the minimum of NLP problem $\min f(x) = (x_1 - 3)^2 + (x_2 - 2)^2$, $g_1(x) = x_1^2 - x_2 - 3 \le 0$, $g_2(x) = x_2 - 1 \le 0$, $g_3(x) = -x_1 \le 0$ with the projected gradient method starting in point $x_0 = (0,0)^T$.

13. Find the minimum of NLP problem $\min f(x) = x_1^2 + x_2^2$, $g(x) = e^{(1-x_1)} - x_2 = 0$ with the sequential quadratic programming approach, starting values $x_0 = (1,0)^T$ and $u_0 = 0$.

6

Deterministic GO algorithms

6.1 Introduction

The main concept of deterministic global optimization methods is that in the generic algorithm description (4.1), the next iterate does not depend on the outcome of a pseudo random variable. Such a method gives a fixed sequence of steps when the algorithm is repeated for the same problem. There is not necessarily a guarantee to reach the optimum solution. Many approaches such as grid search, random function approaches and the use of Sobol numbers are deterministic without giving a guarantee. In Section 6.2 we discuss the deterministic heuristic DIRECT followed by the ideas of stochastic models and response surface methods in Section 6.3. After that we will focus on methods reported in the literature that expose the following characteristics. The method

- solves a problem in a finite number of steps up to a guaranteed accuracy,
- uses the mathematical structure of the problem to be solved.

Example 6.1. Given a problem where we know that feasible set X is a polytope and the objective function f is concave. In Chapter 3 we have seen that the global minimum points must be located in the extreme points. This means one "only" has to consider the vertices to find the optimum. This is called vertex enumeration. One is certain to find the optimum in a finite number of steps. The problem of course is that "finite number of steps" may imply more than a human's lifetime.

The type of deterministic algorithms under consideration is not applicable for black-box optimization problems. Stating the other way around, black-box problems cannot be solved up to an accuracy guarantee. In the literature, usually the argument is given that after evaluating k points, a $k + 1$ degree polynomial can be derived that has a minimum far from the best point found. Alternatively, consider a grid search on the pathological function in Figure 6.1; we are far from the optimum. Structure gives information on how far off we

E.M.T. Hendrix and B.G.-Tóth, *Introduction to Nonlinear and Global Optimization*, 137
Springer Optimization and Its Applications 37, DOI 10.1007/978-0-387-88670-1_6,
© Springer Science+Business Media, LLC 2010

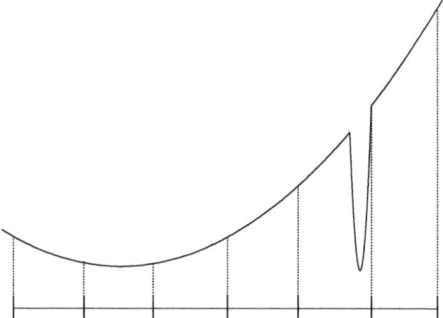

Fig. 6.1. Pathological function; minimum far from best evaluated point

may be. The idea of the discussed deterministic methods is to guarantee the effectiveness of reaching the set of global minimum points. Analysis of effectiveness may tell us that it can take quite some computational effort to reach this target.

Example 6.2. The Piyavskii–Shubert algorithm introduced in Chapter 4 is an example of an algorithm using structure; it requires knowledge of a Lipschitz constant L. It provides a guaranteed distance δ of the function value of the final outcome to the optimum objective function value. Assume that f is constant apart from a small v-shape around the minimum. Then the algorithm requires a full grid over the search interval $[l, r]$. Given a constant L, the algorithm builds a binary tree with $2^D + 1$ trial points, where $D = \lceil (\ln(L \times (r - l)) - \ln \delta)/\ln(2) \rceil - 1$, to reach the guaranteed δ-accuracy. So one knows exactly the finite number of steps.

Deterministic GO methods that aim at a guaranteed accuracy work in a similar way as methods in Combinatorial Optimization. The concepts are based on enumeration, generation of cuts and bounding in such a way that a part of the feasible area is proved not to contain any optimum solution. In branch and bound, it is important to obtain lower bounds of the function f on subregions that are as sharp as possible; the higher a lower bound the better. Section 6.4 describes various important mathematical structures from the literature and discusses how they can be used to derive lower bounds. Section 6.6 illustrates how such bounds can be used in GO branch and bound. In Section 6.7 the concept of cutting planes is illustrated.

6.2 Deterministic heuristic, DIRECT

Heuristics are usually understood as algorithms that provide an approximate solution without any guarantee it is close to the optimal solution. In GO, often heuristics are associated with random search techniques. An easy to understand deterministic heuristic is grid search. Its efficiency is easy to analyze as the number of evaluated points grows exponentially in the dimension. More sophisticated deterministic heuristics are due to random function approaches and radial basis functions discussed in Section 6.3. What is generic in heuristics is that one tries to trade-off local and global search over the feasible area.

The introduction of the DIviding RECTangles algorithm by Jones et al. (1993) and also follow-up literature mention Lipschitz constants and a convex hull. Both concepts are not necessary to describe the basic algorithm. The objective function is not required to be Lipschitz continuous, nor continuous, although it would be nice if it is around the global minimum points. Neither is the idea of dividing rectangles necessary, although it is convenient for the explanation. The algorithm generates a predefined number of N sample points over a grid in a box-constrained feasible area starting from the scaled midpoint $x_1 = \frac{1}{2}(1, 1, \ldots, 1)^T$. The algorithm stays close to the generic description (4.1) in that all sample points $x_1, x_2, \ldots, x_k$ are stored as potential places where refinement may take place. Refinement of x_k consists of sampling more in a region around x_k.

To decide on promising regions where sampling takes place, for each sample point x_k a possibly changing radius vector u_k can be stored to describe the rectangular region $(x_k - u_k, x_k + u_k)$ associated with x_k. Its length $||u_k||$ and function value $f(x_k)$ determine whether x_k is a candidate for refinement. Only one parameter α is used to influence the local versus global trade-off. Three choices describe the algorithm:

- How to select points for refinement.
- How to sample around a chosen point.
- How to update information u_k and the associated rectangle.

We describe each choice, the complete algorithm and illustrate it.

6.2.1 Selection for refinement

The way of sampling over grid points gives that for each iteration a finite number M of sizes of vectors u exist that can be kept ordered $s_1 > s_2 > \cdots > s_M$. Each point and associated u_k falls in a size class S_j corresponding to size s_j, $k \in S_j$. The bi-criterion Figure 6.2 is of importance in the selection.

Algorithm 27 Select subalgorithm $\textbf{select}(f_1, \ldots, f_k, ||u_1||, \ldots, ||u_k||, \alpha)$

Determine $f^U = \min_l f_l$
Sort $||u_1||, \ldots, ||u_k||$ and create classes $S_1, \ldots S_M$ with sizes $s_1 > s_2 > \ldots > s_M$
$m_1 := \min_{k \in S_1} f_k, j := 1$
repeat select $\text{argmin}_{k \in S_j} f_k$
    $j := j + 1$, $m_j := \min_{k \in S_j} f_k$
until $(m_j \geq f^U - \alpha|f^U| + \frac{s_j}{s_{j-1}}(m_{j-1} - f^U + \alpha|f^U|))$

In this figure, for each sample point k, on the x-axis one can find the current rectangle size $\|u_k\|$ and on the y-axis its function value $f(x_k)$. The idea is that by generating new points, current sample points move to smaller sizes, such that the points walk to the left in the figure. The sizes that occur are not equidistant, but as we will see in the way u_k is updated, they contain a certain pattern. One is interested in sampling further around relatively low points ($f(x_k)$ is low) and in relatively unexploited areas ($\|u_k\|$ is big). In a Pareto-like fashion, all nondominated points in the lower right are selected for refinement. At first instance this would mean to select all points that correspond to $m_j = \min_{k \in S_j} f(x_k)$. Here the parameter α is coming in to avoid sampling too locally. On the y-axis, point $f^U - \alpha|f^U|$ is marked, where $f^U = \min_k f(x_k)$ is the record value. This point is added to the so-called nondominated points. A line is drawn from this point upwards such that the resulting curve remains convex. Let us see how this can be done. We start in the class of biggest length S_1 by selecting the point corresponding to minimum m_1 and proceed over the classes j with smaller S_j up to

$$m_j \geq f^U - \alpha|f^U| + \frac{s_j}{s_{j-1}}(m_{j-1} - f^U + \alpha|f^U|), \tag{6.1}$$

such that m_j is higher than the line through $f^U - \alpha|f^U|$ and m_{j-1}. That means that the last lower point(s) are possibly not used for refinement, because the space around it is not empty enough. As stated, this is steered with the only parameter α. In Figure 6.2, $M = 4$ sizes can be distinguished of a vector

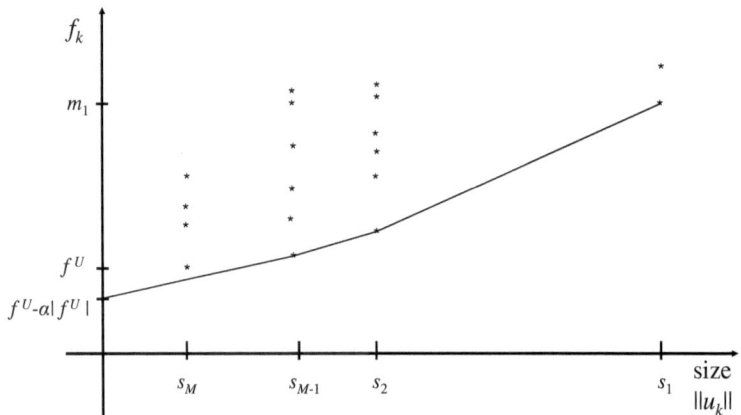

Fig. 6.2. Function value and size of region around sample points

length. The best value $m_M = f^U$ fulfills (6.1) and the corresponding best point found is not selected for refinement. Possibly in further iterations it is, as the selected points move to classes with smaller sizes. Given the function values $f_1, \ldots, f_k$ of the sample points and the sizes of their associated regions

$||u_1||, \ldots, ||u_k||$, we can now determine which sample points are going to be a basis for further sampling. Subalgorithm 27 outlines the procedure. Now that we know where to generate new grid points, the question is how.

6.2.2 Choice for sampling and updating rectangles

The refinement of a point x means that we sample more grid points around it in its hyperrectangle $(x - u, x + u)$. Moreover, the old point x as well as the new sample points get a radius vector assigned smaller in length than u. The sampling around a point x is also steered by the radius vector u. Now not the size is the most important, but the coordinates $i = 1, \ldots, n$ and the corresponding length of the elements u_i. To avoid confusion, we leave out the iteration counter k of the points and focus on the element index i. Due to the process, several coordinates will have the same size of u_i. Therefore, the index set $I = \text{argmax}_i \, u_i$ represents the maximum size edges of the associated hyperrectangle $(x - u, x + u)$. In the DIRECT algorithm this index set plays a big role in determining the new grid points to be evaluated and the way the radius vector is allocated to the points.

The set of new grid points is defined by

$$G = \{x \pm \frac{2}{3} u_i e_i \ \ \forall i \in I\}, \tag{6.2}$$

where e_i is the ith unit vector. They are evaluated and added to the set of sample points. The number of evaluated points grows to $k := k + |G| = k + 2|I|$.

The last item is how to assign radius vectors to the new sample points and how to update the old one. The rule for old refined point x is relatively easy:

$$u_i := \frac{1}{3} u_i, \ \ i \in I, \ \ u_i := u_i, \ \ i \notin I. \tag{6.3}$$

Algorithm 28 Refine subalgorithm **refine(x, u, global k, N)**

Determine $I := \text{argmax}_i \, u_i$	set of maximum sizes of rectangle
for $(i \in I)$ **do**	sample around x
$\quad$ Evaluate $f(x - \frac{2}{3} u_i e_i)$ and $f(x + \frac{2}{3} u_i e_i)$	
$\quad w_i := \min\{f(x - \frac{2}{3} u_i e_i), f(x + \frac{2}{3} u_i e_i)\}$	
$\quad k := k + 2$	
$\quad$ **if** $(k \geq N)$, STOP	
for $(i \in I)$, $v_i := u$	inherit old vector
repeat select $\eta := \text{argmax}_{i \in I} \, w_i$	assign radius vectors u
$\quad$ **for** $(i \in I)$ **do**	
$\quad\quad v_{i\eta} := \frac{1}{3} u_\eta$	reduce size in coordinate η
$\quad\quad u_\eta := \frac{1}{3} u_\eta$	also for original rectangle
$\quad$ Remove η from I	
$\quad$ Store $x_{k-1} := x - \frac{2}{3} u_\eta e_\eta, u_{k-1} := v_\eta$	
$\quad$ Store $x_k := x + \frac{2}{3} u_\eta e_\eta, u_k := v_\eta$	
until $(I = \emptyset)$	

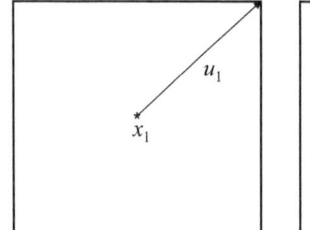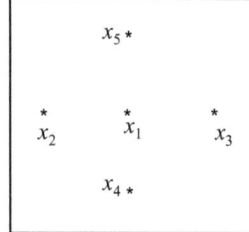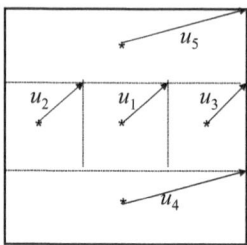

Fig. 6.3. Refinement of x_1, u_1 within DIRECT

For the new points, values $w_i := \min\{f(x - \frac{2}{3}u_i e_i), f(x + \frac{2}{3}u_i e_i)\}$ $\forall i \in I$ are determined. The idea is that new points in coordinate directions with a big value of w_i get a bigger rectangle assigned than the coordinates with lower value of w_i. The lowest w_i coordinate gets the same size rectangle as the old point. The way this can be described is as follows. First the new points inherit the old radius vector u. Then they get a reduction for each coordinate with a higher w_i value. Iteratively find $\eta = \text{argmax}_{i \in I} \, w_i$, give all $i \in I$ a reduction of $\frac{1}{3}$ in the coordinate direction η and remove the new points in coordinate η from the index list I. In this way, the old rectangle is partitioned into new ones around the new and old sample points. The exact pseudocode is given in Algorithm 28. The subalgorithm also has to do global bookkeeping. One should stop sampling when the budget of function evaluations N is exhausted. Moreover, we have to store the new points in the complete list of samples.

Example 6.3. Consider $f(x) = 4x_1^2 - 2.1x_1^4 + \frac{1}{3}x_1^6 + x_1x_2 - 4x_2^4$, the six-hump camel-back function on rectangular feasible area $[-2, 4] \times [-3, 3]$. We will not scale the set, such that $x_1 = (1, 0)$ and $u_1 = (3, 3)$ with $\|u_1\| = 3\sqrt{2}$. As both elements $I = 1, 2$ of u (and sides of the rectangle) are as big, $I = \{1, 2\}$ and four new points are generated; $G = \{(-1, 0), (3, 0), (1, -2), (1, 2)\}$. The new rectangles and corresponding u vectors are determined by $w_1 = f(x_2) = 2.23$ being smaller than $w_2 = f(x_4) = 48.23$. The result of the refinement of x_1, u_1 is depicted in Figure 6.3.

6.2.3 Algorithm and illustration

Now that we have the basic operations of the algorithm, they can be combined to describe the complete DIRECT algorithm. Usually, the instance to be solved is considered to be scaled between 0 and 1 to reduce notation. The algorithm requires storing all evaluated points x_k, their function value f_k and the current radius vector u_k. Notice that in our description the global bookkeeping is done in the refinement step that is keeping hold of the budget N of function evaluations not to be exceeded. The complete algorithm is outlined in Algorithm 29.

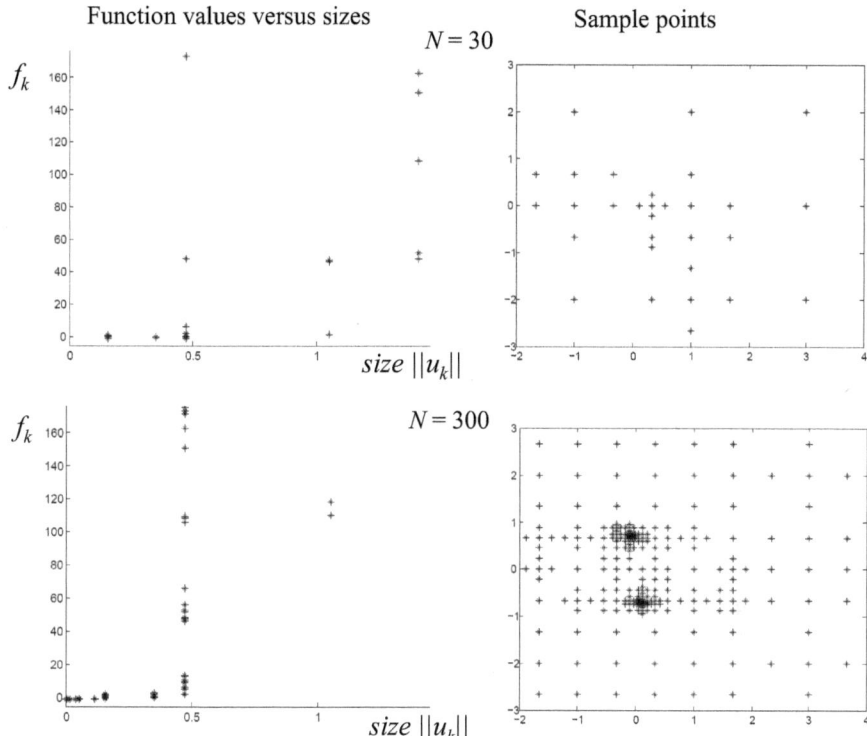

Fig. 6.4. $N = 30$ and $N = 300$ sample points of DIRECT, $\alpha = 10^{-4}$. (Left) the found function values f_k versus the length of the allocated vectors u_k. (Right) the sampled points x_k in the domain. Instance $f(x) = 4x_1^2 - 2.1x_1^4 + \frac{1}{3}x_1^6 + x_1x_2 - 4x_2^4$ on $[-2, 4] \times [-3, 3]$

Example 6.4. We proceed with the six-hump camel-back function of Example 6.3 on feasible area $[-2, 4] \times [-3, 3]$. Algorithm 29 is executed with $\alpha = 10^{-4}$. Resulting sample points after $N = 30$ and $N = 300$ function evaluations are depicted in Figure 6.4. Notice first of all that the graph that gives the function values f_k versus the length of u_k is not as nicely scaled as the illustrative variant in Figure 6.2. Sample points with high function values are only selected

---

Algorithm 29 algorithm **DIRECT(f, α, N)**

$k; = 1, x_1 = \frac{1}{2}(1, 1, \ldots, 1)^T, u_1 = \frac{1}{2}(1, 1, \ldots, 1)^T, f_1 := f(x_1)$
repeat
    $J := \text{select}(f_1, \ldots, f_k, ||u_1||, \ldots, ||u_k||, \alpha)$
    for $(j \in J)$ **do**
        $\text{refine}(x_j, u_j, k, N)$
until (STOP)

for refinement in a later stage of the algorithm. The occurring sizes s_j for two-dimensional instances are either $\sqrt{2}$ or $\sqrt{10}$ times a power of $\frac{1}{3}$. In our run, larger sizes occur, as the feasible set was not scaled between 0 and 1.

Depending on the parameter value for α, more or less sampling is done around points with the lowest function values and less or more sampling is done on a grid in empty areas. In Figure 6.4, the resulting grid is clearly visible. Most of the higher function value points have an associated size of $\|u_k\| = 0.47$, whereas the lower value sample points cluster around the two global minimum points.

Further research was done on the convergence speed of the basic algorithm leading to several suggestions for modifications. Its use was promoted due to MATLAB implementations for research and application. See, e.g., Finkel and Kelley (2006); Björkman and Holmström (1999).

6.3 Stochastic models and response surfaces

As function evaluations may be time consuming, many researchers have been studying how to best use the information of the evaluated points in order to generate a most promising next sample point x_{k+1}. The idea to use a stochastic model to select the next point to be evaluated is usually attributed to Kushner (1962). Given the evaluated sample points p_i and their function value y_i, $i = 1, \ldots, k$, the objective function is modeled as a random variable $\xi_k(x)$. Essential is that in fact the model ξ_k should coincide with the evaluated function values, so $P(\xi_k(p_i) = y_i) = 1$, $i = 1, \ldots, k$. The next point x_{k+1} to be evaluated is based on maximizing for instance an expected utility value $U(\cdot)$:

$$x_{k+1} = \operatorname*{argmax}_x E[U(\xi_k(x))]. \tag{6.4}$$

Also the term random function approach has been used. Although the terms stochastic and random are used, the resulting algorithms are basically deterministic as no random effect is used to generate x_{k+1}. The numerous possibilities to define ξ_k and the choice of the criterion in (6.4), lead to follow-up research in what we would now call the Lithuanian and Russian school. Mockus dedicated a book to the approach, Mockus (1988), and Antanas Žilinskas elaborated many variants as explained in his books Törn and Žilinskas (1989) and Zhigljavsky and Žilinskas (2008). He investigated the properties of what is called the P-algorithm.

Let $f^U = \min_i y_i$ be the best function value found thus far and δ_k be a kind of positive aspiration level on improvement in iteration k. The P-algorithm takes as the next iterate the point where the probability of reaching a value below $f^U - \delta_k$ is maximum:

$$x_{k+1} = \operatorname*{argmax}_x P(\xi_k(x) < f^U - \delta_k). \tag{6.5}$$

The ease of solving (6.5) depends on the construction of the stochastic model. To get a feeling, it is to be noted that if $\boldsymbol{\xi}_k$ is Gaussian, we capture the model into the mean $m_k(x)$ and variance $s_k^2(x)$ and write (6.5) as the equivalent

$$x_{k+1} = \operatorname*{argmax}_x \frac{f^U - \delta_k - m_k(x)}{s_k(x)}. \tag{6.6}$$

Notice that $s_k(p_i) = 0$, as $\boldsymbol{\xi}_k$ is assumed to be known in these points.

The idea of the random function is close to what in spatial statistics is called "Kriging"; see Ripley (1981). Such models are based on interpolating measurements. As such the concept of interpolation or response surface modeling is not the most appealing. Consider that we would fit a surface through the measurements of Figure 6.1. The minimum of any fitted curve would not be close to the peak. Recently, several papers generated more interest in random function approaches by linking them to response surfaces and radial basis functions. First, the paper of Jones et al. (1998) linked the work of many researchers and the idea of response surfaces with stochastic models. Second, the paper of Gutmann (2001) elaborated the radial basis function interpolation ideas of Mike Powell with its use in Global Optimization. It was directly recognized that the concepts are close to random function approaches. Later Žilinskas showed the equivalence with the P-algorithm. Due to the implementation in software and the application to practical problems, the concepts became more known.

The idea of radial basis functions is to interpolate f at x given the function values y_i at p_i by taking values more into account when they are closer, i.e., $r = \|x - p_i\|$ is smaller. This is done by using a so-called radial basis function, for instance $\theta(r) = \exp(-r^2)$. Now define

$$\varphi_k(x) = \sum_i w_i \theta(\|x - p_i\|), \tag{6.7}$$

where the k weights w_i can be determined by equating $\varphi_k(p_j) = y_j$, i.e., by solving $w = \Theta(p)^{-1} y^T$. The entrances of matrix $\Theta(p)$ are given by $\Theta_{ij}(p) = \theta(\|p_i - p_j\|)$. Notice that if the iterations k proceed, this matrix is growing. As we are dealing with an interpolation type of description, the minimum of φ_k will tend to the best point found.

Hans-Martin Gutmann elaborated many variants, and we now describe an idea close to the earlier sketched P-algorithm. In that sense we should define an aspiration level $f^U - \delta$. However, instead of thinking in stochastic models or maximum likelihood, the terminology is that of "bumpiness" and a so-called seminorm $w^T \Theta(p) w = y \Theta^{-1}(p) y^T$ to be minimized. This means that a point x is chosen for x_{k+1}, such that adding $(x, f^u - \delta)$ to p, y should be "most appropriate" in terms of minimizing the seminorm. Notice that if $x = p_i$, then the matrix $\Theta(p, x)$ is singular and the seminorm infinite. Writing this in one expression,

$$x_{k+1} = \min_x(y, f^U - \delta)\Theta^{-1}(p, x)(y, f^U - \delta)^T. \tag{6.8}$$

Adding x_{k+1} to the measurement points $p_{k+1}, y_{k+1} = f(x_{k+1})$ defines a special case of an algorithm that uses the radial basis function as response surface to find the global minimum. We illustrate with an example.

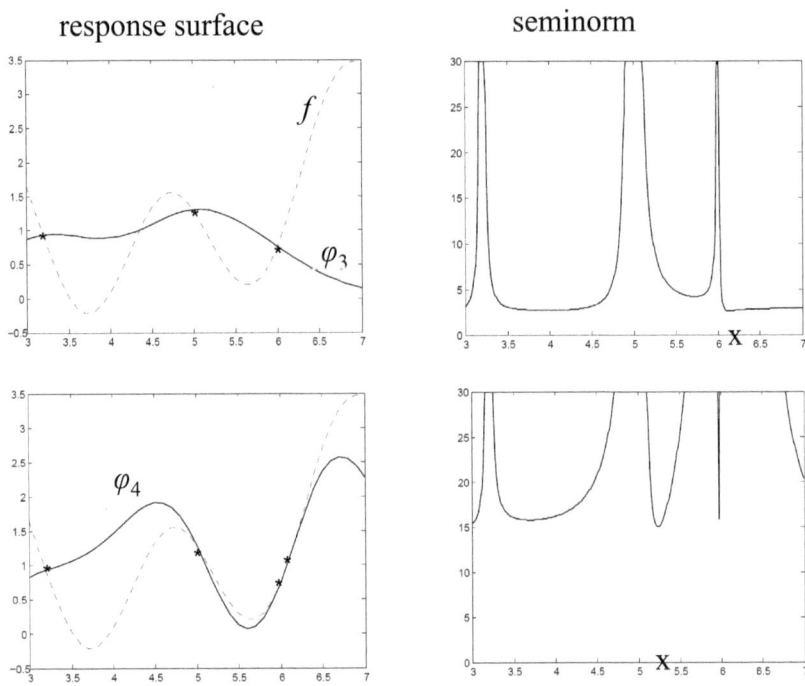

Fig. 6.5. Response surface based on three and four observations in the left of the graph. Corresponding seminorm (6.8) on the right. Its minimum point x is used to select the next sample point for the response surface

Example 6.5. Consider function $f(x) = \sin(x) + \sin(3x) + \ln(x)$ on interval $[3, 7]$. Three points have been evaluated, namely, $p_1 = 3.2, p_2 = 5$ and $p_3 = 6$. In Figure 6.5 we first see the corresponding graph of response surface $\varphi_3(x)$ in (6.7) based on $\theta(r) = \exp(-r^2)$. By using $f^U = 0.76 = f(6)$ and choosing $\delta = 0.1$ now the seminorm of (6.8) is defined. Its graph is given in the upper right of Figure 6.5. One can see that at the observation points p_i it goes to infinity. Its minimum is used as next measurement point $p_4 = x_{k+1} = 6.12$ resulting in the improved response surface $\varphi_4(x)$. The corresponding seminorm has a minimum point in $x_5 = 5.24$.

The idea of the random function approach and the use of the seminorm is that the function evaluations are so expensive that they outweigh the increasing computation we have to do due to taking information of all evaluated points into account; the matrix $\Theta(p)$ gets bigger and bigger. The example shows how in order to find the following evaluation point, we have to solve another global optimization problem.

6.4 Mathematical structures

The literature on deterministic Global Optimization contains lots of analysis on mathematical structures, such as quadratic, concave, bilinear structures that can be used for the construction of specific algorithms. An overview is provided by Horst and Tuy (1990). It goes too far to mention all of the literature which appeared in the field. Nevertheless, it is worthwhile to mention the appearance of another summary, Horst and Pardalos (1995), which includes more than deterministic approaches only. Further monographs are appearing.

We focus on how structural information can be obtained and used. Moreover, a new aspect is that we distinguish two types of structures, A and B.

A. Analysis necessary to reveal structure
This class of structures contains among others:

- Concavity
- d.c.: difference of convex functions
- Lipschitz continuity

Although not always easy to verify, concavity has the strong advantage that no further value information is required to exploit this structure. On the contrary, nearly every practical objective function is as well d.c. as Lipschitz continuous. However, the use of the structures requires a so-called d.c. decomposition for the first and a so-called Lipschitz constant for the second. This will be illustrated.

B. Mathematical expression reveals structure
The following structures are used:

- Quadratic functions
- Bilinear functions
- Multiplicative functions
- Fractional functions
- Interval arithmetic on explicit expressions

The classes are not mutually exclusive, but mainly a view to approach the function to be minimized or bounded (cut). One structure can be translated to another. The main objective is to derive bounds on a set. A so-called *minorant* of f on X is a function φ such that $\varphi(x) \leq f(x)$, $\forall x \in X$. A *convex envelope* is specifically a minorant which is convex and sharp, i.e., there is no better one: there does not exist a convex g with $\varphi(x) \leq g(x) \leq f(x), \forall x \in X$ and $\exists x \in X, \quad g(x) > \varphi(x)$). Minorants and convex envelopes are used to derive bounds.

6.4.1 Concavity

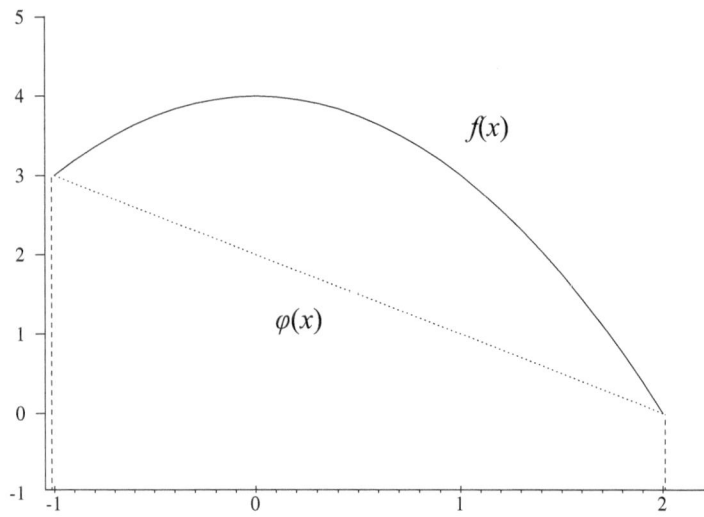

Fig. 6.6. Concave function f and affine minorant φ

Often, the term nonconvex optimization is related to global optimization. This refers directly to Theorem 3.10: If f is a convex function on convex set X, there is only (at most) one local and global minimum. The most common structure of multiextremal problems is therefore nonconvexity. On the other hand, minimizing a nonconvex objective function f, does not necessarily imply the existence of multiple optima but may explain the occurrence. Concavity can be called an extreme form of nonconvexity. A property used by deterministic methods is related to Theorem 3.11: If f is a concave function on compact X, the local minimum points coincide with extreme points of X.

Example 6.6. Given concave function $f(x) = 4 - x^2$ on feasible set $X = [-1, 2]$. The extreme points of the interval are the minimum points; see Figure 6.6.

When in general the feasible set X is a polytope, then in a worst case situation, every vertex may correspond to a local minimum. Some algorithms are based on performing an efficient so-called vertex enumeration.

Minimizing a concave objective function on a closed convex feasible set is called concave programming. Figure 6.6 also shows the possibility of constructing a so-called affine underestimating function $\varphi(x)$, based on the definition of concave functions. Given two iterates x_k and x_l and their corresponding function values $f_k = f(x_k)$ and $f_l = f(x_l)$, the function value for every convex combination of the iterates, $x = \lambda x_k + (1 - \lambda)x_l$, is underestimated by

$$f(x) = f(\lambda x_k + (1 - \lambda)x_l) \geq \lambda f_k + (1 - \lambda)f_l = \varphi(x), \quad 0 \leq \lambda \leq 1. \quad (6.9)$$

Example 6.7. For Example 6.6 this works as follows. Let $x_k = 2$ and $x_l = -1$. An arbitrary point x in $[-1, 2]$ is a convex combination of the extreme points: $x = \lambda x_k + (1 - \lambda)x_l = 2\lambda - (1 - \lambda) \rightarrow \lambda = (x + 1)/3$. Now affine function $\varphi(x) = \lambda f(2) + (1 - \lambda)f(-1) = 3(1 - \lambda) = 2 - x$ underestimates $f(x)$ on $[-1, 2]$.

The minorant $\varphi(x)$ can be used to derive lower bounds of the minimum objective function value on a bounded set. We illustrate its use in branch and bound in Section 6.6.1 and for cutting planes in Section 6.7.

Concavity of the objective function from a given practical model formulation may be hard to identify. Concavity occurs for instance in situations of economies of scale. Following Theorem 3.8, in cases where f is two times differentiable one could check whether the eigenvalues of the Hessean are all nonpositive. The eigenvalues, representing the second derivatives, give a measure how concave the function is. Notice that the affine underestimator $\varphi(x)$ does not require the value information of the eigenvalues. This is a strong point of the structure. Notice furthermore that the underestimation becomes worse, less tight, when f is more concave, the second-order derivatives are more negative.

6.4.2 Difference of convex functions, d.c.

Often, a function can be written as the difference of two convex functions, $f(x) = f_1(x) - f_2(x)$. For the function $f(x) = 3/(6 + 4x) - x^2$ in Figure 6.7, which has two minima on the interval $[-1, 1]$, this is easy to see. Splitting the function in a difference of two convex functions is called a d.c. decomposition. For the example function, a logical choice is to consider $f(x)$ as the difference of $f_1(x) = 3/(6 + 4x)$ and $f_2(x) = x^2$. The construction of a convex underestimating function of f proceeds as follows. The concave part $-f_2(x)$ is underestimated by an affine underestimating function $\varphi_2(x)$ based on (6.9) and added to the convex part f_1. In this way a convex underestimating function $f_1 + \varphi_2$ appears, which can be used to derive lower bounds of the objective function on bounded sets.

Example 6.8. For the function $f(x) = 3/(6 + 4x) - x^2$, $\varphi_2(x) = -1$ underestimates $-f_2(x) = -x^2$ resulting in the convex minorant $\varphi_{dc1} = 3/(6 + 4x) - 1$ in Figure 6.7. The decomposition is not unique. Often in the literature, the argument is used that the second derivative may be bounded below; in this case for instance by a value of -8. A decomposition can be constructed by adding a convex function with a second derivative of 8 and subtracting it again: $f_1(x) = f(x) + 4x^2$ and $f_2(x) = 4x^2$. The resulting convex minorant $\varphi_{dc2} = f(x) + 4x^2 - 4$ depicted in Figure 6.7, is less tight than the first one.

This example teaches us several things. Indeed, nearly every function can be written as d.c. by adding and subtracting a strong convex function. The condition that the second derivative is bounded below is sufficient. For practical

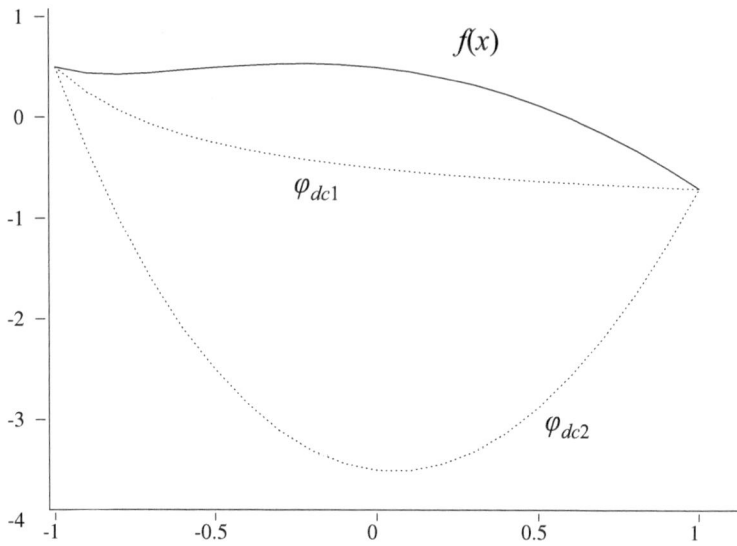

Fig. 6.7. Two convex minorants of a d.c. function

algorithmic development, first a d.c. decomposition has to be constructed. If the lower bound on the second derivative is used, value information is necessary.

Another related structure is the so-called concept of reverse convex programming, i.e., minimizing a convex function on a convex set intersected by one reverse convex constraint; the constraint defines the complement of a convex set. A d.c. described function on a convex set X can be transformed to a reverse convex structure by defining the problem:

$$\min z + f_1(x), z \geq -f_2(x), \quad x \in X. \tag{6.10}$$

The dimension of the problem increases by one, as the variable z is added.

Example 6.9. In Example 6.8, the transformation may lead to reverse convex program $\min z + 3/(6 + 4x), z \geq -x^2, x \in [-1, 1]$.

Both structures require the same type of approaches. For further theoretical results on d.c. programming we refer to the overview by Tuy (1995).

6.4.3 Lipschitz continuity and bounds on derivatives

In nearly every practical case, the function f to be optimized is also so-called Lipschitz continuous. Practically it means that its slope is bounded on the feasible region X. More formally, there exists a scalar L such that

$$\mid f(x_1) - f(x_2) \mid \leq L\|x_1 - x_2\| \qquad \forall x_1, x_2 \in X. \tag{6.11}$$

Mainly due to the one-dimensional algorithms of Shubert (1972) and Danilin and Piyavskii (1967), Lipschitz continuity became well known in Global Optimization. The validation whether a function is Lipschitz continuous is, in contrast to concavity, not very hard. As long as discontinuities, or "infinite derivatives" do not occur, e.g., when f is smooth on X, the function is also Lipschitz continuous. The relation with derivatives (slopes) is given by

$$L \geq \frac{|f(x_1) - f(x_2)|}{\|x_1 - x_2\|} \qquad \forall x_1, x_2 \in X. \tag{6.12}$$

For differentiable instances L can be estimated by

$$L = \max_{x \in X} \|\nabla f(x)\|. \tag{6.13}$$

The requirement of value information is more obvious using Lipschitz continuity in algorithms than using d.c. decomposition. Notice that (6.11) also applies for any overestimate of the Lipschitz constant L. Finding such a guaranteed overestimate is in general as difficult as the original optimization problem. In test functions illustrating the performance of Lipschitz optimization algorithms, trigonometric functions are often used so that estimates of the Lipschitz constant can be derived easily. As illustrated by sawtooth cover Figure 4.5 and Algorithm 4, the guarantee not to miss the global optimum is based on the lower bounding

$$f(x) \geq f_k - L\|x - x_k\| \tag{6.14}$$

at iteration points x_k. Although the Piyavskii–Shubert algorithm was formulated for one-dimensional functions, it stimulated many multidimensional elaborations. It is interesting from a geometric perspective to combine the cones (6.14) of all iterates x_k, f_k. A description of the corresponding algorithm is given by Mladineo (1986). One can observe various approximations in the literature to deal with the multivariate problem seen from a branch and bound perspective. Among others, Meewella and Mayne (1988) change the norm and obtain lower bounds on subsets via Linear Programming. In Pintér (1988) and Sergeyev (2000), one can observe approaches that focus on the diagonal of box-shaped regions.

Another interesting direction is due to the work of Breiman and Cutler (1993). Their focus is on a bound K on the second derivative, such that $-K \leq f''(x), x \in X$ or more general (in higher dimensions) an overestimate of the negative of the minimum eigenvalue of the Hessean. The analogy of (6.14) is given by

$$f(x) \geq f_k + f_k'(x - x_k) - \frac{1}{2}K(x - x_k)^2 \tag{6.15}$$

for a function of one variable and in general

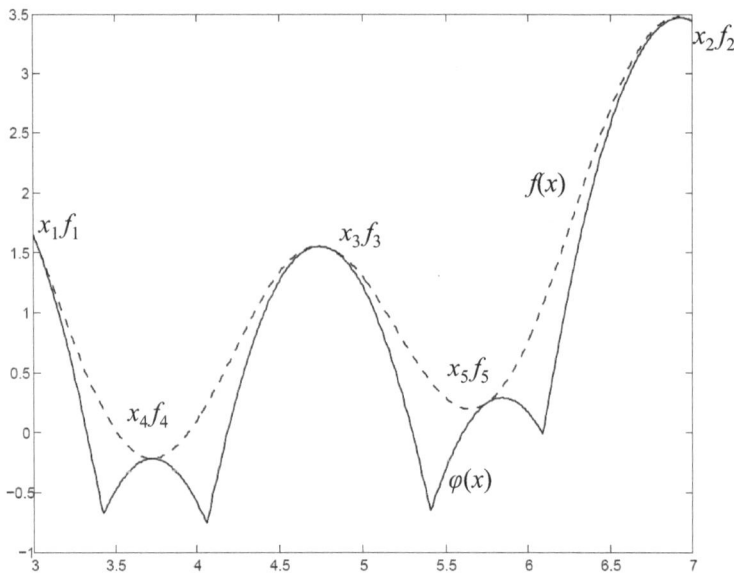

Fig. 6.8. Breiman–Cutler algorithm for $f(x) = \sin(x) + \sin(3x) + \ln(x)$ given $K = 10$

$$f(x) \geq f_k + \nabla f_k^T(x - x_k) - \frac{1}{2}K\|x - x_k\|^2. \qquad (6.16)$$

Now the underestimating function $\varphi(x)$ can be taken as the maximum over k of the parabolas (6.16). The algorithm of Breiman–Cutler takes iteratively a minimum point of φ as next iterate.

Example 6.10. The absolute value of the second derivative of $f(x) = \sin(x) + \sin(3x) + \ln(x)$ is bounded by 10 on the interval $[3, 7]$; $K = 10$. Figure 6.8 depicts what happens if we take for the next iterate a minimum point of $\varphi(x) = \max_k\{f_k + f_k'(x - x_k) - \frac{1}{2}K(x - x_k)^2\}$.

In more-dimensional cases this leads to interesting geometric structures. Bill Baritompa studied additions on cutting away regions where the optimum cannot be. Several articles (e.g., Baritompa, 1993) show what he called the "southern hemisphere" view that it is not necessary to have a global overestimate of either Lipschitz constant L or second derivative K. Knowing the local behavior around the global minimum point x^* is sufficient and usually better to cut away larger areas. Let K^* be a value

$$f(x) \leq f^* + \frac{1}{2}K^*\|x - x^*\|^2, \ \forall x. \qquad (6.17)$$

Given an iterate x_k, f_k (6.17) tells us about the optimum x^*, f^* that $f^* \geq f_k - \frac{1}{2}K^*\|x_k - x^*\|^2$. This means that the area under

$$\varphi(x) = \max_k\{f_k - \frac{1}{2}K^*\|x - x_k\|^2\} \qquad (6.18)$$

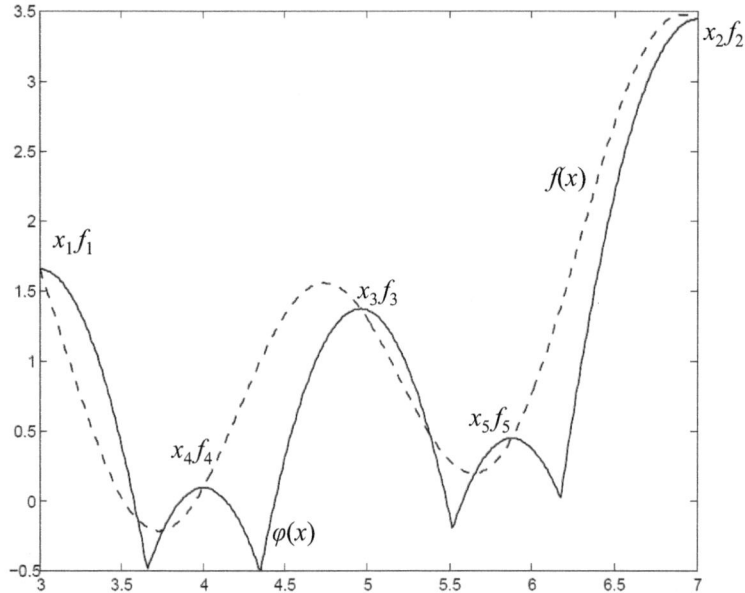

Fig. 6.9. Iterate is a minimum of (6.18) for $f(x) = \sin(x) + \sin(3x) + \ln(x)$

cannot contain the global minimum. The interesting aspect is that φ is not necessarily an underestimating function of f.

Example 6.11. For function $f(x) = \sin(x) + \sin(3x) + \ln(x)$ we take $K^* = 10$, as the maximum value of the second derivative is taken close to x^*. Iteratively taking a minimum point of (6.18) does not result in a lower bounding function, but neither cuts away the global minimum. The minimum point of φ is a lower bound for the minimum of f. The process is illustrated in Figure 6.9.

The same reasoning applies for

$$L^* = \max_{x \in X} \frac{|\, f(x) - f(x^*)\,|}{\|x - x^*\|} \tag{6.19}$$

where the optimum cannot be below the sawtooth "cover" defined by

$$\varphi(x) = \max_k \{f_k - L^* \|x - x_k\|\}. \tag{6.20}$$

The sawtooth cover with slope L^* is not necessarily an underestimating function everywhere, but neither cuts away the global minimum.

Example 6.12. For the function $f(x) = \sin(x) + \sin(3x) + \ln(x)$ on the interval $X = [3, 7]$ the maximum $L^* = 2.67$ of (6.19) is attained for boundary point $x = 3$ as sketched in Figure 6.10. Taking $L^* = 2.67$ in the Piyavskii–Shubert algorithm gives that the tooth of the first iteration directly runs

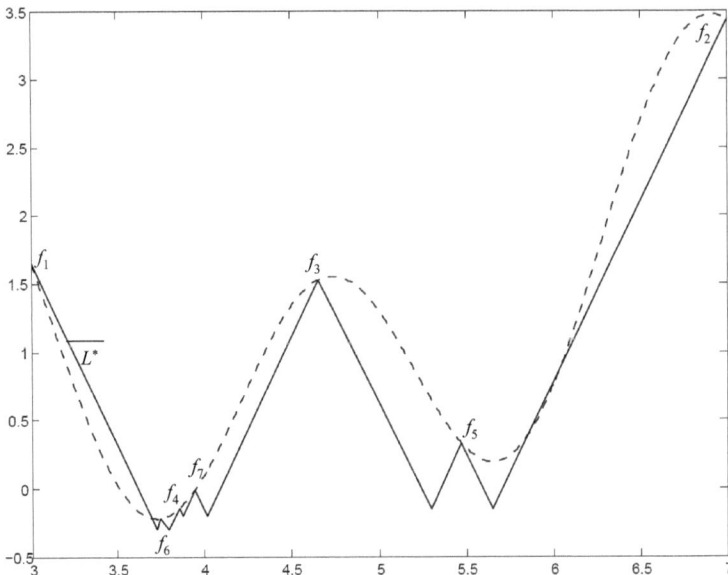

Fig. 6.10. Piyavskii–Shubert for $f(x) = \sin(x) + \sin(3x) + \ln(x)$ given $L^* = 2.67$

through the minimum, which is approached very fast. Iterate $x_6 = 3.76$ approaches minimum point $x^* = 3.73$ very close and in the end, only the intervals $[x_1, x_6] = [3, 3.76]$ and $[x_6, x_4] = [3.76, 3.86]$ are not discarded. Figure 6.10 shows that the sawtooth cover does not provide a minorant, although in the end we have the guarantee to enclose the global minimum point.

The use of structures is very elegant as such, as illustrated here. On the other hand, practically value information is required. We now discuss the second type of structures, where such information is relatively easy to obtain.

6.4.4 Quadratic functions

Quadratic functions have a wide applicability in economics and regression. Also in the literature on mathematical optimization they get a lot of attention. As introduced in Chapter 3 equation (3.16), they can be written as

$$f(x) = x^T A x + b^T x + c. \tag{6.21}$$

It is not difficult to recognize quadratic functions in a given model structure, as only linear terms and products of two decision variables occur in the model description. The matrix A is a symmetric matrix which defines the convexity of the function in each direction. Eigenvectors corresponding to positive eigenvalues define the directions in which f is convex, negative eigenvalues give the concavity (negative second derivatives) of the function in the direction of the corresponding eigenvectors. Depending on the occurrence of positive and negative eigenvalues of A, the function can be either concave (all eigenvalues negative), convex (all positive) or indefinite (positive as well as negative eigenvalues).

If the function is concave, corresponding affine underestimation can be used, as will be illustrated in Section 6.6. The eigenvalues in that case give an indication as to the quality of the underestimation; more negative values give a less tight underestimation. The d.c. view can easily be elaborated by splitting the function in a convex and concave part due to a so-called eigenvalue decomposition. Also the value information for the Lipschitzian view can be found relatively easy. As the derivative $2Ax + b$ is linear, the length $\|2Ax + b\|$ is convex, such that its maximum can be found in one of the extreme points of a subset under consideration. A bound on the second derivative can be directly extracted from the eigenvalues of A.

In quadratic programming problems, f is minimized on a polyhedral set X. Due to the linearity of the derivatives, the Karush–Kuhn–Tucker conditions for the local optima are a special case of the so-called Linear Complementarity Problem, which is often discussed in the optimization literature. For a further overview, see Horst et al. (1995).

Example 6.13. Consider $f(x) = 3x_1^2 - 3x_2^2 + 8x_1x_2$ from Example 3.8; $A = \begin{pmatrix} 3 & 4 \\ 4 & -3 \end{pmatrix}$ and $b = 0$. A Lipschitz constant over a bounded set can be found by maximizing $\|2Ax + b\|^2 = 4x^T A^T Ax = 100x_1^2 + 100x_2^2$, which is a convex function. The eigenvalues of A are $\mu_1 = 5$ and $\mu_2 = -5$ with eigenvectors $r_1 = \frac{1}{\sqrt{5}}(2,1)^T$ and $r_2 = \frac{1}{\sqrt{5}}(1,-2)^T$. An eigenvalue decomposition lets A be written as

$$A = \frac{1}{5} \begin{pmatrix} 2 & 1 \\ 1 & -2 \end{pmatrix} \begin{pmatrix} 5 & 0 \\ 0 & -5 \end{pmatrix} \begin{pmatrix} 2 & 1 \\ 1 & -2 \end{pmatrix}. \tag{6.22}$$

From a d.c. decomposition point of view this means that now $f(x) = x^T Ax$ can be written as $f(x) = f_1(x) - f_2(x)$ by taking

$$f_1(x) = x^T \frac{1}{5} \begin{pmatrix} 2 & 1 \\ 1 & -2 \end{pmatrix} \begin{pmatrix} 5 & 0 \\ 0 & 0 \end{pmatrix} \begin{pmatrix} 2 & 1 \\ 1 & -2 \end{pmatrix} x = 4x_1^2 + x_2^2 + 4x_1x_2$$

and

$$f_2(x) = x^T \frac{1}{5} \begin{pmatrix} 2 & 1 \\ 1 & -2 \end{pmatrix} \begin{pmatrix} 0 & 0 \\ 0 & 5 \end{pmatrix} \begin{pmatrix} 2 & 1 \\ 1 & -2 \end{pmatrix} x = x_1^2 + 4x_2^2 - 4x_1x_2.$$

6.4.5 Bilinear functions

For bilinear functions, the vector of decision variables can be partitioned into two groups (x, y) and the function can be written in the form

$$f(x, y) = c^T x + x^T Qy + d^T y, \tag{6.23}$$

in which Q is not necessarily a square matrix. The function is linear whenever either the decision variables x or the decision variables y are fixed. Actually, bi-affine would be a better name, as the function becomes affine in one group of variables when the other group is fixed. The roots of bilinear programming can be found in Nash (1951), who introduced game problems involving two players. Each player must select a mixed strategy from fixed sets of strategies open to each, given knowledge of the payoff based on selected strategies. These problems can be treated by solving a so-called bilinear program. Bilinear problems are interesting from a research point of view, because of the numerous applied problems that can be formulated as bilinear programs; such as dynamic Markovian assignment problems, multicommodity network flow problems, quadratic concave minimization problems. For an overview, we refer to Al-Khayyal (1992). One of the properties is that the optimum is attained at the boundary of the feasible set.

The underestimation is based on so-called Linear Programming relaxations. The basic observation is that for a product of variables xy on a box $x \in [l_x, u_x]$ and $y \in [l_y, u_y]$

$$\begin{aligned} xy &\geq l_x y + l_y x - l_x l_y \\ xy &\geq u_x y + u_y x - u_x u_y. \end{aligned} \qquad (6.24)$$

Example 6.14. Consider the function $f(x, y) = -2x - y + xy$ on the box constraint $0 \leq x \leq 4$ and $0 \leq y \leq 3$. The function has two minima attained in the vertices $(0, 3)^T$ and $(4, 0)^T$. Elaboration of (6.24) gives

$$\begin{aligned} xy &\geq 0 \\ xy &\geq 3x + 4y - 12. \end{aligned}$$

So the function $\varphi(x, y) = \max\{0, 3x+4y-12\}$ is a minorant of xy on $0 \leq x \leq 4$ and $0 \leq y \leq 3$. Function $f(x, y)$ can be underestimated by $-2x - y + \varphi(x, y)$.

The use of the minorant will be illustrated in Section 6.6.

6.4.6 Multiplicative and fractional functions

A function is called a multiplicative function when it consists of a multiplication of convex functions. Besides the multiplication of two variables, as in bilinear programming, higher-order terms may occur. A multiplicative function consists of a product of several affine or convex functions. It may not be hard to recognize this structure in a practical model formulation. For an overview on the mathematical properties we refer to Konno and Kuno (1995).

A function f is called fractional or rational when it can be written as one ratio or the sum of several ratios of two functions, $f(x) = \frac{g(x)}{h(x)}$. The ratio of two affine functions got most attention in the literature. Depending on the structure of the functions g and h, the terminology of linear fractional programming, quadratic fractional programming and concave fractional programming is applied.

A basic property is due to Dinkelbach (1967). Let the function $\theta(x)$ be defined as $\theta(x) = \{g(x) - \lambda h(x)\}$. If the (global) minimum λ^* of $f(x)$ is used as the parameter in the function $\theta(x)$, then the minimum point of θ (with objective value zero) corresponds to a global minimum point of f. This property can be used for bounding in the following way. Let f^U correspond to a found objective function value. Then we can ask ourselves whether a subset X of the feasible area contains a better (lower) function value than f^U:

$$\min_X \left\{ f(x) = \frac{g(x)}{h(x)} \right\} \leq f^U,$$

which translates into

$$\min_X \{g(x) - f^U h(x)\} \leq 0. \tag{6.25}$$

If the latter is not the case, one does not have to consider subset X anymore. For an overview on fractional programming, we refer to Schaible (1995).

Example 6.15. Consider the function $f(x) = \frac{g(x)}{h(x)}$, where g is a convex quadratic function $g(x) = 2x_1^2 + x_2^2 - 2x_1 x_2 - 6x_1 + 1$ and h is linear $h(x) = 2x_1 - x_2 + 0.1$. One can imagine it is convenient that h does not become zero on the domain which we take as the box constrained area $X = [3, 6] \times [0, 6]$.

Consider first an upper bound $f^U = -3$. The question whether better values can be found on X is given by (6.25) such that we want to minimize $g(x) + 3h(x) = 2x_1^2 + x_2^2 - 2x_1 x_2 - 3x_2 + 1.3$ over X. It can be shown that the unique minimum can be found at $x = (3, 4.5)^T$ with an objective function value of -0.95. So indeed better function values can be found.

We now consider the Dinkelbach result from the perspective of maximizing $f(x)$. While function g has a global maximum attained in $(6, 0)^T$, the fractional function f has a global maximum of 10 in $(3, 6)^T$ and two more local optima. Defining

$$\theta_1(x) = \{g(x) - 10h(x)\} = 2x_1^2 + x_2^2 - 2x_1 x_2 - 26x_1 + 10x_2$$

shows more clearly that we are maximizing a convex quadratic function. One can verify that it has a global maximum of 0 in $(3, 6)^T$ and no local nonglobal maxima. Now we focus on the minimum. One can show for instance by using a solver that f has a minimum of about -3.66 in the boundary point $(3, 4.8)$. Writing the equivalent

$$\theta_2(x) = \{g(x) + 3.66h(x)\} = 2x_1^2 + x_2^2 - 2x_1 x_2 + 1.33x_1 - 3.66x_2 + 1.37$$

gives a quadratic function with a minimum on the feasible area of 0 at $(3, 4.8)$. One can verify it fulfills the Karush–Kuhn–Tucker conditions.

6.4.7 Interval arithmetic

The concepts of interval arithmetic became known due to the work of Moore (1966) with a focus on error analysis in computational operations. The use for Global Optimization has been elaborated in Hansen (1992) and Kearfott (1996). The main concepts are that of thinking in boxes (interval extensions) and inclusion functions. $\mathbb{I} = \{X = [a, b] \mid a \leq b; a, b \in \mathbb{R}\}$ is the set of the one-dimensional intervals. Then $X = [\underline{x}, \overline{x}] \in \mathbb{I}$ is a one-dimensional interval which extends the idea of a real x. A box $X = (X_1, \ldots, X_n)$, $X_i \in \mathbb{I}$, $i = 1, \ldots, n$, is an n-dimensional interval as element of $\mathbb{I}^n$. Where the range of an element $w(X_i) = (\overline{x}_i - \underline{x}_i)$ is given, the width $w(X) = \max_{i=1,\ldots,n} w(X_i)$ is defined as a kind of accuracy. Let $f(X) = \{f(x) \mid x \in X\}$ be the real range of f on X, then F and $F' = (F'_1, \ldots, F'_n)$ are called interval extensions of f and its derivatives ∇f. The word inclusion is used to express that $f(X) \subseteq F(X)$ and $\nabla f(X) \subseteq F'(X)$.

An inclusion function generally overestimates the range of a function. The extent of the overestimation depends on the type of the inclusion function, on the considered function and on the width of the interval. If the computational costs are the same, the smaller the overestimation, the better is the inclusion function.

The main idea behind interval analysis is the natural extension of real arithmetical operations to interval operations. For a pair of intervals $X, Y \in \mathbb{I}$ and arithmetical operator $\circ \in \{+, -, \cdot, /\}$ one extends to $X \circ Y = \{x \circ y \mid x \in X, y \in Y\}$. That is, the result of the interval operation contains all the possible values obtained by the real operation on all pairs of values belonging to the argument intervals. Because of the continuity of the operations, these sets are intervals. For the division operation, zero should not belong to the denominator. As arithmetic operations are monotonous, the definitions of the corresponding interval versions are straightforward:

$$X + Y = [\underline{x} + \underline{y}, \overline{x} + \overline{y}] \tag{6.26}$$

$$X - Y = [\underline{x} - \overline{y}, \overline{x} - \underline{y}] \tag{6.27}$$

$$X \cdot Y = [\min\{\underline{x}\underline{y}, \overline{x}\underline{y}, \underline{x}\overline{y}, \overline{x}\overline{y}\}, \max\{\underline{x}\underline{y}, \overline{x}\underline{y}, \underline{x}\overline{y}, \overline{x}\overline{y}\}]$$

$$\frac{1}{Y} = \left[\frac{1}{\overline{y}}, \frac{1}{\underline{y}}\right], \quad 0 \notin Y. \tag{6.28}$$

Example 6.16. Many textbooks emphasize that $x^2 - x = x(x - 1)$, but

$$[0, 1]^2 - [0, 1] = [0, 1] - [0, 1] = [-1, 1] \quad \text{and}$$
$$[0, 1]([0, 1] - 1) = [0, 1][-1, 0] = [-1, 0].$$

Among others, Kearfott (1996) shows that bounds are sharp, which in that context means that the bounds coincide with minimum and maximum, if terms appear only once:

For $f(x) = x^2 - 2$, $F[-2, 2] = [-2, 2]^2 - 2 = [0, 4] - [2, 2] = [-2, 2]$.
For $f(x_1, x_2) = x_1 x_2$, $F([-1, 1], [-1, 1])^T = [-1, 1][-1, 1] = [-1, 1]$.

For monotonic functions the interval extension is relatively easy. For instance,

$$\sqrt{X} = [\sqrt{\underline{x}}, \sqrt{\overline{x}}], \ X \geq 0 \qquad (6.29)$$
$$\ln(X) = [\ln(\underline{x}), \ln(\overline{x})], \ X > 0 \qquad (6.30)$$
$$e^X = [e^{\underline{x}}, e^{\overline{x}}]. \qquad (6.31)$$

For a nonmonotonic function, such as sine or cosine, its periodic nature can be used to construct its interval extension. For use in branch and bound methods, the essential idea is that $w(F(X)) \rightarrow 0$ if $w(X) \rightarrow 0$.

6.5 Global Optimization branch and bound

We first sketch the basic form of branch and bound (B&B) methods and then elaborate it for several illustrative cases in the following sections. The basic idea in B&B methods consists of a recursive decomposition of the original problem into smaller disjoint subproblems until the solution is found. The method avoids visiting those subproblems which are known not to contain a solution. B&B methods can be characterized by four rules: *Branching, Selection, Bounding,* and *Elimination* (Ibaraki, 1976; Mitten, 1970). For problems where the solution is determined when a desired accuracy is reached, a *Termination rule* has to be incorporated.

Algorithm 30 sketches a generic scheme for cases where lower bound calculation also involves the generation of a feasible point. It generates approximations of global minimum points that are less than δ in function value from the optimum. The method starts with a set C_1 enclosing the feasible set X of the optimization problem. For simplicity, we assume minimization and the set X to be compact. At every iteration the branch and bound method has a list Λ of subsets (partition sets) C_k of C_1. In GO, several geometric shapes are used for that like cones, boxes and simplices. The method starts with C_1 as the first element and stops when the list is empty. For every set C_k in Λ, a lower bound f_k^L of the minimum objective function value on C_k is determined. For this, the mathematical structures discussed in Section 6.4 are used.

At every stage, there also exists a global upper bound f^U of the minimum objective function value over the total feasible set defined by the objective value of the best feasible solution found thus far. The bounding (pruning) operation concerns the deletion of all sets C_k in the list with $f_k^L > f^U$, also called cut-off test. Besides this rule for deleting subsets from list Λ, a subset can be removed when it does not contain a feasible solution. In Algorithm 30, index r represents the number of subsets which have been generated. Note that r does not give the number of subsets on the list.

There are several reasons to remove subsets C_k from the list, or alternatively, not to put them on the list in the first place.

- C_k cannot contain any feasible solution.

Algorithm 30 Outline branch and bound algorithm $\mathbf{B \ \& \ B}(X, f, \delta, \epsilon)$

Determine a set C_1 enclosing feasible set X, $X \subset C_1$,
Determine a lower bound f_1^L on C_1 and a feasible point $x_1 \in C_1 \cap X$
if (there exists no feasible point) STOP
else $f^U := f(x_1)$; Store C_1 in Λ; $r := 1$
while $(\Lambda \neq \emptyset)$
    Remove (selection rule) a subset C from Λ and split it
        into h new subsets $C_{r+1}, C_{r+2}, \ldots, C_{r+h}$
    Determine lower bounds $f_{r+1}^L, f_{r+2}^L, \ldots, f_{r+h}^L$
    for $(p := r + 1 \text{ to } r + h)$ **do**
        if $(C_p \cap X$ contains no feasible point)
            $f_p^L := \infty$
      if $(f_p^L < f^U)$
            determine a feasible point x_p and $f_p := f(x_p)$
            if $(f_p < f^U)$
                $f^U := f_p$
                remove all C_k from Λ with $f_k^L > f^U$ cut-off test
            if $(f_p^L > f^U - \delta)$
                Save x_p as an approximation of the optimum
            else if $(Size(C_p) \geq \epsilon)$ store C_p in Λ
    $r := r + h$
endwhile

- C_k cannot contain the optimal solution as $f_k^L > f^U$.
- C_k has been selected to be split.
- It has no use to split C_k any more. This may happen when the size $Size(C_k)$ of the partition set has become smaller than a predefined accuracy ϵ, where

$$Size(C) = \max_{v, w \in C} \|v - w\|. \tag{6.32}$$

Branching concerns further refinement of the partition. This means that one of the subsets is selected to be split into new subsets. There exist several ways for doing so. The selection rule determines the subset to be split next, and influences the performance of the algorithm. One can select the subset with the lowest value for its lower bound (best first search) or for instance the subset with the largest size (relatively unexploited); breadth first search. The target is to obtain sharp bounds f^U soon, such that large parts of the search tree (of domain C_1) can be pruned.

Specific interval B&B algorithms keep the concept of enclosing the optimum by a box. The final result is a list of boxes whose union certainly contains the global optimizers and not a list of global optimum points. The upper bound is not necessarily a result of evaluating the function value in a point, but can be based on the lowest (guaranteed) upper bound over all boxes. Moreover, in differentiable cases there is also a so-called monotonicity test. One focuses on partial derivatives $\frac{\partial f}{\partial x_j}(x) = \nabla_j f(x)$, which for a box C_k

is included by $F'_j(C_k)$. If $0 \notin F'_j(C_k)$, it means it cannot contain a stationary point and one should consider whether C_{kj} contains a boundary point of the feasible area. If this is not the case, one can discard the box.

After a successful search, list Λ will be empty and a guarantee is given either that the global optimum points have been found or that there exists no feasible solution. However, in practical situations, the size of list Λ may keep increasing and filling up the available computer memory, despite possible use of efficient data structures. In the following, we illustrate the B&B procedure for several specific cases.

6.6 Examples from nonconvex quadratic programming

We specify Algorithm 30 for the quadratic programming problem making use of the structures discussed in Section 6.4. The algorithms should find all global optimum points for the general (nonconvex) quadratic programming problem on a compact set; i.e., $f(x) = x^T A x + b^T x + c$ and X is a polytope.

As partition sets (hyper)rectangles (boxes) C_k are used, defined by the two extreme corners l and u, i.e., $l_{jk} \leq x_j \leq u_{jk}, j = 1, \ldots, n$. Initially the global upper bound f^U can be set to infinity or given the objective function value of a feasible solution of X which can be found by LP. Two ways of calculating a lower bound f^L are elaborated and used for small numerical examples.

Lower bound "concave" The lower bound based on concave minimization can be applied when $f(x)$ is concave. Let the vertices of box C_k be v_i, $i = 1, \ldots, 2^n$, with corresponding function values $F_i = f(v_i)$. We define the convex piecewise affine underestimating function $\varphi_k(x) =$

$$
\begin{aligned}
\min_{\lambda} \quad & \sum_{i=1}^{2^n} F_i \lambda_i && \text{(convex combination function values)} \\
\text{s.t.} \quad & \sum_{i=1}^{2^n} v_i \lambda_i = x && \text{(convex combination vertices)} \\
& \sum_{i=1}^{2^n} \lambda_i = 1 && \text{(weights)} \\
& \lambda_i \geq 0 \quad i = 1, \ldots, 2^n.
\end{aligned}
\tag{6.33}
$$

The difference between φ_k and f becomes automatically smaller when the partition set C_k becomes small. The lower bound f^L_k can now be found by solving LP problem (6.33) minimizing over x as well as λ:

$$
f^L_k = \min_{x \in X} \varphi_k(x).
\tag{6.34}
$$

Notice that due to equations (6.33), in (6.34) we implicitly minimize over C_k.

Lower bound "bilinear" A lower bound based on the equivalence of quadratic programming with bilinear programming (Al-Khayyal, 1992) can be applied. The quadratic term $x^T A x$ is equivalent to

$$x^T y, \quad y = Ax. \tag{6.35}$$

Now the bilinear inequality (6.24) can be used, if we have easily available bounds $[l^y, u^y]$ on $y = Ax$ given a box $C = [l^x, u^x]$ on x. This is indeed the case. First define an indicator I that selects the upper and lower bound of $C = [l^x, u^x]$ depending on the sign of a matrix element a:

$$I(C, j, a) = \begin{cases} l_j^x & \text{if } a < 0, \\ u_j^x & \text{if } a \geq 0. \end{cases} \tag{6.36}$$

Based on this indicator we can construct upper and lower bounds for y:

$$\begin{aligned} l_i^y &= \sum_j a_{ij} I(C, j, a_{ij}), & i = 1, \dots, n \\ u_i^y &= \sum_j a_{ij} I(C, j, -a_{ij}), & i = 1, \dots, n. \end{aligned} \tag{6.37}$$

Given rectangle $C = [l^x, u^x]$ with corresponding bounds $[l^y, u^y]$ for $y = Ax$, the lower bound is based on solving LP problem

$$\begin{aligned} f_k^L = \min \ & \sum_j \alpha_j + b^T x + c \\ \text{s.t. } \ & y = Ax, \\ & \alpha_j \geq l_j^x y_j + l_j^y x_j - l_j^x l_j^y, \quad j = 1, \dots, n \\ & \alpha_j \geq u_j^x y_j + u_j^y x_j - u_j^x u_j^y, \quad j = 1, \dots, n \\ & x \in X \cap C_k. \end{aligned} \tag{6.38}$$

Now the question is what it looks like, if we put the branch and bound algorithm in practice with these lower bounds. For the illustration we elaborate two examples.

6.6.1 Example concave quadratic programming

Example 6.17. Consider the following concave quadratic program:

$$\min_{x \in X} \{ f(x) = 4 - (x_1 - 1)^2 - x_2^2 \}. \tag{6.39}$$

$X = \{ 0 \leq x_1 \leq 2.5, 0 \leq x_2 \leq 2, -x_1 + 8x_2 \leq 11, x_1 + 4x_2 \leq 7, 6x_1 + 4x_2 \leq 17 \}$. Contour lines and feasible area are depicted in Figure 6.11. The problem has four local optimum points; $(0, 1.375)^T, (1, 1.5)^T, (2, 1.25)^T$ and $(2.5, 0.5)^T$. Moreover, points $(0, 0)^T, (1, 0)^T$ and $(2.5, 0)^T$ are Karush–Kuhn–Tucker points which are not local optima.

To run the branch and bound algorithm, many choices have to be made. First of all, the choice of the partition sets. We use boxes (hyperrectangles) with as first set $C_1 = [l^x, u^x] = [(0, 0)^T, (2.5, 2)^T]$. The choice of the splitting

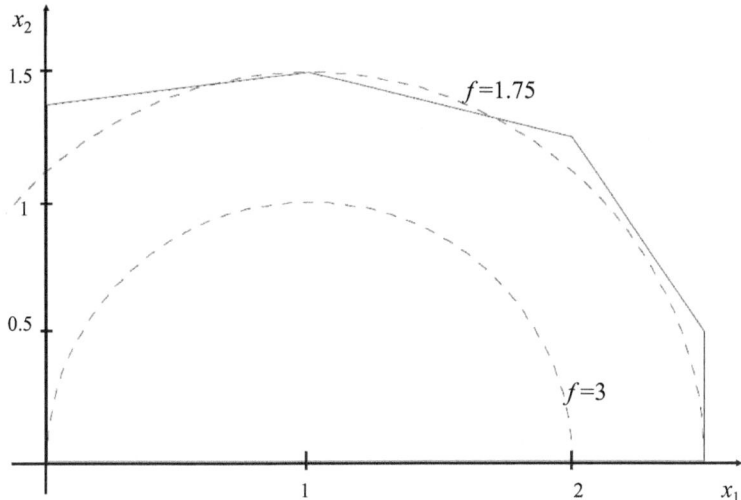

Fig. 6.11. Example concave quadratic program

is to bisect them over the longest edge into two new subsets. For each box C_k, a lower bound is determined by solving problem (6.33), which also provides x_k. Its function value is used to update global upper bound f^U. The selection criterion is of major importance for the course of the algorithm; which subset is selected to be split next? In this example, the subset with the lowest lower bound is selected. For the final accuracy we take here a value of $\delta = 0.05$. The resulting course of the algorithm is given in Figure 6.12.

Example 6.17 shows several generic aspects. One can observe that the global optimum $x^* = (0, 1.375)$ is found in an early stage of the algorithm. Actually, it is even an alternative solution to x_1 of problem (6.33). The further iterations only serve as a verification of the optimality of x^*. It is important to find a good global bound soon. During the course of this specific algorithm we observe that x_k often is the same point. However, the difference for the subsets is that if we get deeper into the tree, its lower bound is going up. Mainly what is necessary for convergence is that the gap between lowest lower bound and upper bound f^U is closing. Implicitly, an accuracy δ determines the stopping. One can observe that C_{10}, which encloses the global optimum, like C_1, C_2, C_7 and C_8, is not split further because $f^L > f^U - \delta$. This means that we have the guarantee to be closer than $\delta = 0.05$ in objective function value from the global optimum. The bounding leads to removing subsets C_4, C_6, C_9, C_{12} and C_{13} because $f^L > f^U$. Subset 11 appeared to be infeasible.

6.6.2 Example indefinite quadratic programming

Example 6.18. Consider indefinite quadratic problem (2.2), where we minimize

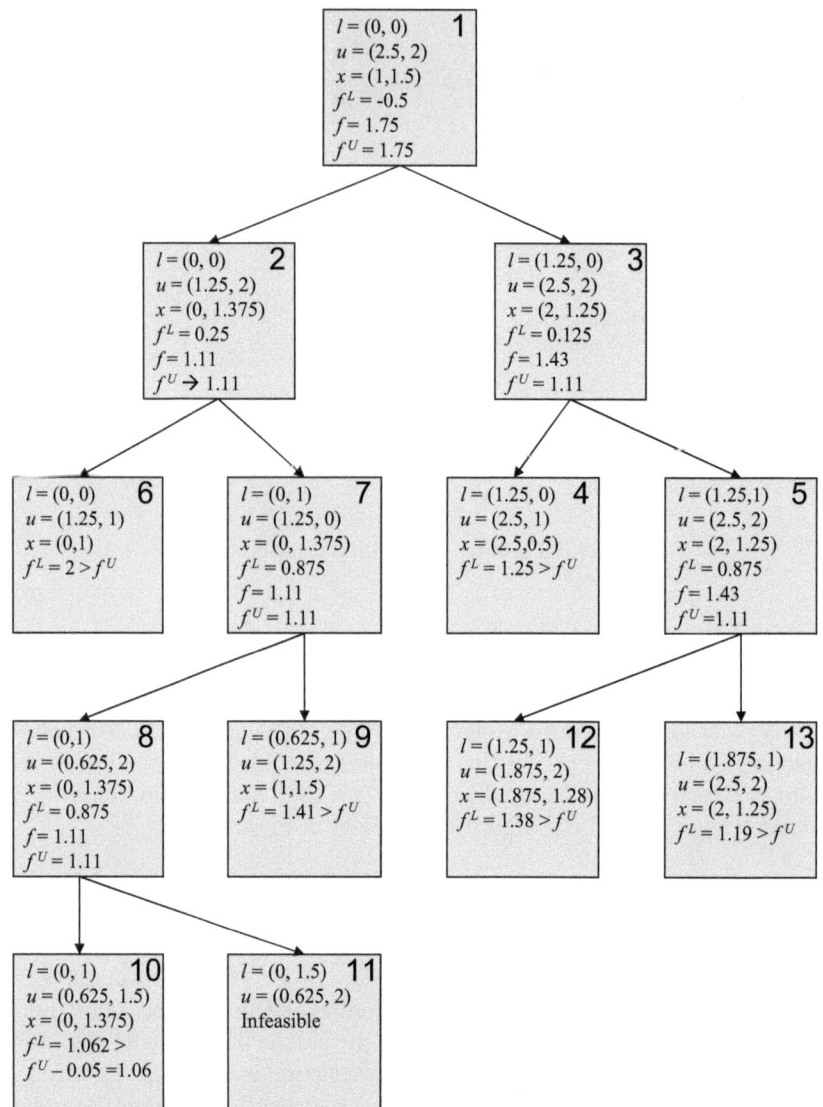

Fig. 6.12. Resulting B&B tree for concave optimization

$$\min\{f(x) = (x_1 - 1)^2 - (x_2 - 1)^2\} \tag{6.40}$$

over $X = \{0 \le x_1 \le 3, 0 \le x_2 \le 4, x_1 - x_2 \le 1, 4x_1 - x_2 \ge -2\}$. Contour lines and feasible area are depicted in Figure 2.12. The problem has local optima in the points $(1, 0)^T$ and $(1, 4)^T$. The first set is logically $C_1 = [l^x, u^x] = [(0, 0)^T, (3, 4)^T]$. For each box C_k, the lower bound calculation is now based on the bilinear concept by solving problem (6.38). The calculation of the bounds for the y variable is extremely simple as $y_1 = x_1$ and $y_2 = -x_2$.

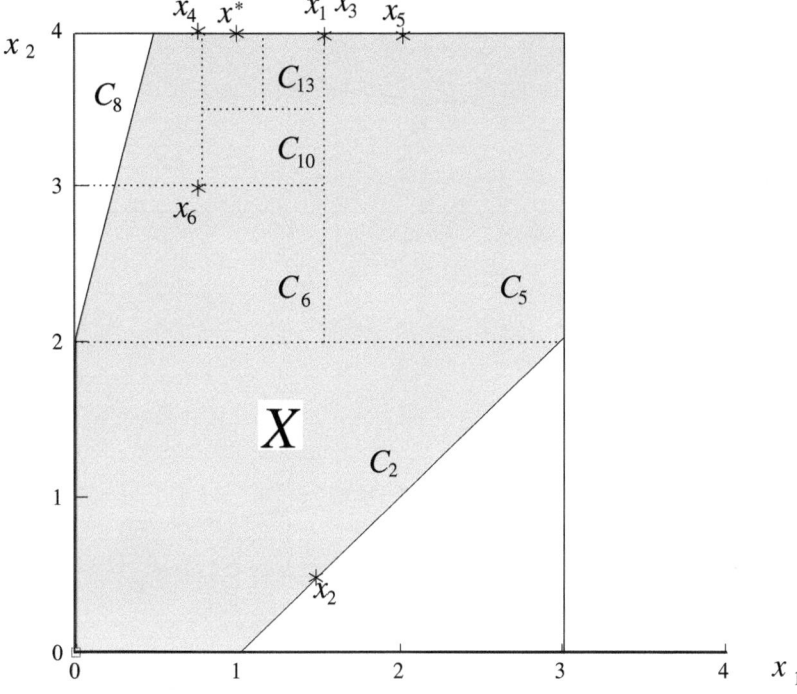

Fig. 6.13. Final partition running B&B

For the final accuracy we take here a value of $\delta = 0.05$. The course of the algorithm is given in Figure 6.14. The final partition is depicted in Figure 6.13. The selection criterion is in this case not relevant, as the algorithm behaves like a local search leaving at most one subset in list Λ. In an early stage already large parts of the feasible set can be removed, e.g., C_2. One observes several improvements of the best point found. Finally, $x_{12} = (0.94, 4)^T$ is taken as an approximation of the global optimum and the algorithms stops, as $f^U - \delta < f^L_{12} < f^U$ where no subsets are left on the list. Once more during the iterations it stops exploring similarly, when it finds $f^L_8 > f^U - \delta$, so that x_8 is temporarily saved as an estimation of the optimum.

Typical in this example is that the lower bound does not always improve, despite the fact that the subset is getting smaller. Observe for instance the shrinking of subset C_1 to C_3 and C_9 to C_{11}. Theoretically, the algorithm will always converge due to a final check on the size of the remaining box. Practically, the designer of an algorithm feels more comfortable when the gap between lowest lower bound and upper bound is shrinking with a guaranteed step size.

6.7 Cutting planes

In general, cutting planes (hyperplanes) serve to discard parts of the search region. In mixed integer programming so-called Gomory cuts are cutting planes,

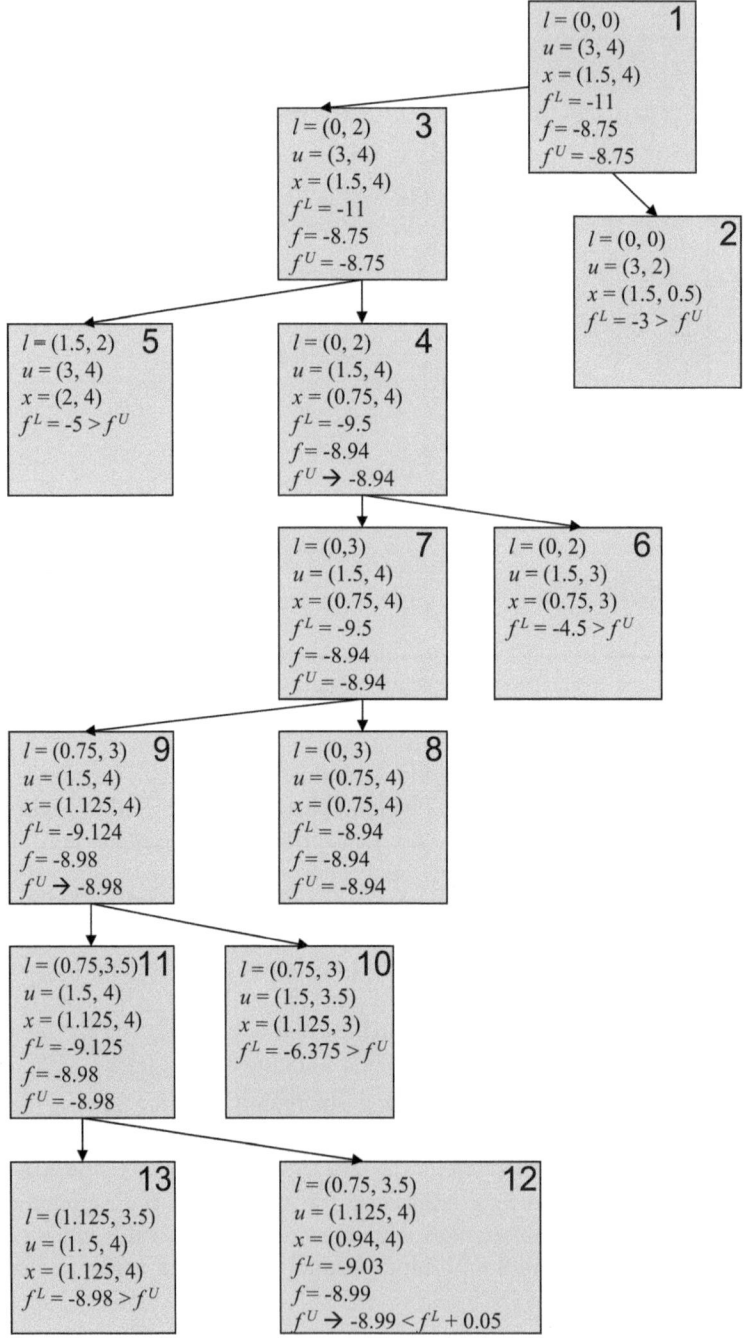

Fig. 6.14. Resulting B&B tree for indefinite example

and in convex optimization Kelley's method works with cutting planes; see Kelley (1999). They are important in the development of large-scale optimization algorithms where NLP and integer programming are mixed.

Here we introduce the concept of cutting planes focusing on concave programming. In concave optimization we want to minimize a concave function over a convex set, typically given by linear constraints. As we have discussed in Chapter 3, the minimum points can be found at a vertex (or vertices) of the feasible region.

To calculate a cutting plane, consider a vertex v of the convex set together with its neighbors $v_1, v_2, \ldots, v_n$, and the best function value $\gamma = f^U$ found so far. Starting from v, one determines the largest step size in the direction of each neighbor, for which the function value is larger than γ. To be precise, denote by $d_i = v_i - v$ the direction toward neighbor v_i. Now for $i = 1, \ldots, n$ determine θ_i such that

$$f(v + \rho_i d_i) \geq \gamma \qquad \forall \; 0 \leq \rho_i \leq \theta_i.$$

The term γ-extension is used in this context. The n points $(v + \theta_i d_i)$ define a hyperplane, that can be considered as cutting plane. One can cut off the part of the feasible set where v lies, because there is no point with better function value than $\gamma = f^U$. Occasionally, we may cut off a point x with $f(x) = \gamma$, but that is not a problem as we already know that point.

Cutting planes can be generated until the convex set disappears, such that one stops where the minimum f^* coincides with $\gamma = f^U$. Instead of formalizing an algorithm, we have a look at what such a procedure would look like for the instance in Example 6.17.

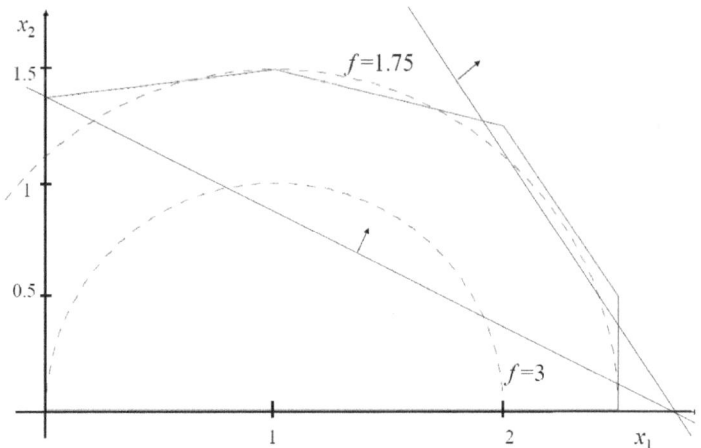

Fig. 6.15. Cutting planes for Example 6.19

Example 6.19. Consider the concave quadratic problem of Example 6.17, i.e.,

$$\min_{x \in X}\{f(x) = 4 - (x_1 - 1)^2 - x_2^2\}. \tag{6.41}$$

$X = \{0 \le x_1 \le 2.5, 0 \le x_2 \le 2, -x_1 + 8x_2 \le 11, x_1 + 4x_2 \le 7, 6x_1 + 4x_2 \le 17\}$.
Let $v = (0,0)^T$, so the neighbors are $v_1 = (0, 1.375)^T, v_2 = (2.5, 0)^T$ with
objective function values $f(v) = 4, f(v_1) = 1.11, f(v_2) = 1.75$ so $\gamma = 1.11$. To
determine θ_1, consider $f(v + \rho_1 d_1)$, where $d_1 = v_1 - v = (0, 1.375)^T$; $f(v + \rho_1 d_1) = 4 - (1.375\rho_1)^2 - 1$, such that $\theta_1 = 1$. Similarly, $d_2 = v_2 - v = (2.5, 0)^T$
and $f(v + \rho_2 d_2) = 4 - (2.5\rho_2 - 1)^2$ from which $\theta_2 = 1.08$. The two points which
define the cutting plane are $(0, 1.375)^T$ and $(2.7, 0)^T$. The corresponding cut
can be described as $(1.375, 2.7)^T x \ge 3.78$ or $x_1 + 1.96x_2 \ge 2.7$.

Consider $v = (0, 1.375)^T$ as next iterate. The new neighbors are $v_1 = (1, 1.5)^T, v_2 = (2.5, 0.1)^T$ with objective function values $f(v) = 1.11, f(v_1) = 1.5, f(v_2) = 1.74$ so γ is still 1.11. Now, $d_1 = (1, 0.125)^T$ and $d_2 = (2.5, -1.273)^T$. To determine θ_1, the function $f(v + \rho_1 d_1) = 4 - (\rho_1 - 1)^2 - (1.375 + 0.125\rho_1)^2$ is considered, while for θ_2 the inequality $f(v + \rho_2 d_2) = 4 - (2.5\rho_2 - 1)^2 - (1.375 - 1.273\rho_2)^2 \ge \gamma$ must hold. One can verify that
$\theta_1 = 1.63$ and $\theta_2 = 1.08$. Thus, the points of the cutting plane are $(1.63, 1.58)^T$
and $(2.7, 0)^T$, and the cutting plane is $(1.58, 1.07)^T x \ge 4.27$.

The problem with the two cutting planes is depicted in Figure 6.15.

6.8 Summary and discussion points

- After a given number of function evaluations no guarantee can be given
  on the distance from the best point found to a global optimum point, if
  no structural information is used.
- Heuristic methods can be used to handle problems with expensive function
  evaluations. DIRECT and stochastic model algorithms store and use information of all evaluated points. Moreover, the sketched algorithm based on
  radial basis functions requires solving a global optimization problem for
  choosing the next point to be evaluated.
- Knowing concavity requires no further value information to construct a
  rigorous method.
- Mathematical structures of d.c. (difference of convex functions), Lipschitz
  continuity and bounds on higher derivatives need value information to be
  applied in an algorithmic context.
- Methods applying quadratic, bilinear, fractional programming or interval
  arithmetic contain procedures to generate necessary information for solving problems up to a guaranteed accuracy.
- With respect to efficiency, it is not known if the latter effectiveness is
  reached within a human's lifetime.
- Algorithms based on branch and bound contain choices like selection rule
  and priority of tests that influence the efficiency.

- The B&B methods require effort to be implemented and a design of efficient data structures, such that the memory can still be handled.

6.9 Exercises

1. Given function $f(x) = 1 - 0.5x + \ln(x)$. Show that f is concave on the interval $[1, 4]$. Give an affine minorant $\varphi(x)$ of f on $[1, 4]$.

2. Given function $f(x) = -x_1^3 + 3x_1^2 + x_2$ on the feasible set $X = [-1, 1] \times [0, 2]$. Determine the Lipschitz constant of f on X. Give the lower bounding expression (6.14) given $x_k = (0, 1)^T$.

3. Given function $f(x) = -x_1^3 + 3x_1^2 + x_2$ on the feasible set $X = [-1, 1] \times [0, 2]$. Determine K as the maximum absolute eigenvalue of the Hessean of f over X. Give the lower bounding expression (6.16) given $x_k = (0, 1)^T$.

4. Given function $f(x) = x_1 x_2$. Find a d.c. decomposition of f, i.e., write $f(x) = f_1(x) - f_2(x)$ such that f_1, f_2 are convex. How would you construct a convex underestimating function $\varphi(x)$ based on the decomposition over feasible set $X = [-1, 1] \times [-1, 2]$?

5. Given function $f(x) = x_1 x_2$ on feasible set $X = [-1, 1] \times [-1, 2]$. Give an underestimating function $\varphi(x_1, x_2)$ considering the bilinear nature of the function. Determine a lower bound of f by minimizing $\varphi(x)$ over X.

6. Given fractional function $f(x) = \frac{x_1 + x_2}{2x_1 + 3x_2 + 1}$ on feasible set $X = [1, 2] \times [1, 2]$. Given a bound $f^U = 0.5$. Determine whether f has lower values than f^U on X. Determine the global minimum λ^* of f on X by analyzing the sign of the partial derivatives. Minimize $(x_1 + x_2) - \lambda^*(2x_1 + 3x_2 + 1)$ over X.

7. Given function $f(x) = (x_1 + x_2)^2 - |x_1|$ on feasible interval $X = [-1, 2] \times [1, 2]$. Generate a lower and upper bound (inclusion) of f over X based on the natural interval extension. Consider the two subintervals that appear when we split the longest side of X over the middle. Determine now the bounds for these two new intervals. Which of them would you split further first when minimizing f?

8. Given a value $K > f''(x), \forall x \in X$ according to (6.18) one can determine a bound z_p on the optimum f^* for an interval $[l_p, r_p]$. Elaboration gives

$$z_p = \frac{f(l_p) + f(r_p)}{2} - \frac{1}{2K}\left(\frac{f(l_p) - f(r_p)}{r_p - l_p}\right)^2 - \frac{K}{8}(r_p - l_p)^2. \qquad (6.42)$$

The minimum point of φ on $[l_p, r_p]$ is given by

$$m_p = \frac{1}{K}\frac{f(l_p) - f(r_p)}{r_p - l_p} + \frac{r_p + l_p}{2}. \qquad (6.43)$$

A possible branch and bound algorithm close to Piyavskii–Shubert is given in Algorithm 31. Construct a branch and bound tree following this algorithm for finding the minimum of $f(x) = \sin(x) + \sin(3x) + \ln(x)$ on the

---

Algorithm 31 Barit($[l, r], f, K, \delta$)

Set $p := 1$, $l_1 := l$ and $r_1 := r$, $\Lambda = \{[l_1, r_1]\}$

$f^U := \min\{f(l), f(r)\}$, $x^U := \mathrm{argmin}\{f(l), f(r)\}$

$z_1 := \frac{f(l) + f(r)}{2} - \frac{1}{2K}(\frac{f(l) - f(r)}{r - l})^2 - \frac{K}{8}(r - l)^2$

while ($\Lambda \neq \emptyset$)

    remove an interval $[l_k, r_k]$ from Λ with $z_k = \min_p z_p$

    evaluate $f(m_k) := f(\frac{1}{K}\frac{f(l_k) - f(r_k)}{r_k - l_k} + \frac{r_k + l_k}{2})$

    if ($f(m_k) < f^U$)

        $f^U := f(m_k)$, $x^U := m_k$ and remove all C_p from Λ with $z_p > f^U - \delta$

    split $[l_k, r_k]$ into 2 new intervals $C_{p+1} := [l_k, m_k]$ and $C_{p+2} := [m_k, r_k]$

        with corresponding lower bounds z_{p+1} and z_{p+2}

    if ($z_{p+1} < f^U - \delta$) store C_{p+1} in Λ

    if ($z_{p+2} < f^U - \delta$) store C_{p+2} in Λ

    $p := p + 2$

endwhile

---

interval $X = [3, 7]$. Take $K = 10$ and for the accuracy use $\delta = 0.005$. Compare your tree with Figure 4.6.

7

Stochastic GO algorithms

7.1 Introduction

We consider stochastic methods as those algorithms that use (pseudo) random numbers in the generation of new trial points. The algorithms are used a lot in applications. Compared to deterministic methods they are often easy to implement. On the other hand, for many applied algorithms no theoretical background is given that the algorithm is effective and converges to a global optimum. Furthermore, we still do not know very well how fast the algorithms converge. For the effectiveness question, Törn and Žilinskas (1989) already stress that one should sample "everywhere dense". This concept is as difficult with increasing dimension as doing a simple grid search. In Section 7.2 we describe some observations that have been found by several researchers on the question of increasing dimensions.

For the efficiency question, the literature is often looking at the process from a Markovian perspective (Zhigljavsky and Žilinskas, 2008). This means that the probability distribution for the next trial point depends on a certain state that has been reached. This view allows analysis in the convergence speed. For practical algorithms, often it is impossible to distinguish a state space and stationary process for the Markovian view. Theoretical results can be derived by not looking at implemented algorithms, but by investigating ideal algorithms. In Section 7.4 we will sketch the idea of Pure Adaptive Search and the analysis.

According to the generic description of Törn and Žilinskas (1989):

$$x_{k+1} = Alg(x_k, x_{k-1}, \ldots, x_0, \xi), \tag{7.1}$$

where ξ is a random variable, a random element enters the algorithm. One perspective is to consider ξ as a stationary random variable. Another way is to say that the algorithm is adapting the distribution of ξ in each iteration. For the analysis this distinction is important. It is also good to realize that in practice we are not dealing with random numbers, but with pseudo-random

E.M.T. Hendrix and B.G.-Tóth, *Introduction to Nonlinear and Global Optimization*, 171
Springer Optimization and Its Applications 37, DOI 10.1007/978-0-387-88670-1_7,
© Springer Science+Business Media, LLC 2010

numbers generated by a computer. This also gives rise to all kinds of practical variants such as the use of so-called Sobol numbers to get a more uniform cover of the space, or using stratified sampling. Initially most analysis focused on Pure Random Search (PRS), Multistart and when to stop sampling. Development of effective variants followed where one applies clustering with the former two strategies. We sketch the concept of clustering in Section 7.3.

Summarizing, there is a difference in the popularity of applying stochastic algorithms and the difficulty of deriving scientific theoretical results such as summarized in Boender and Romeijn (1995) and Zhigljavsky and Žilinskas (2008). In our personal experience cooperating in engineering applications, we feel that there is a belief that stochastic algorithms solve optimization problems, whereas we only know that they generate many candidates of which one selects the best one. This belief seems to be fed by physical and biological analogies. Although population algorithms for GO were already in existence, they really became popular due to a wave of so-called genetic algorithms that appeared in the seventies. The evolutionary analogy is as attractive as the idea of simulated annealing, particle swarms, ant colony analogies, etc. In Section 7.5, several of these algorithms are described.

7.2 Random sampling in higher dimensions

The final target of algorithms is to come close to the global optimum, or the set of global optimum solutions. When the number of decision variables grows, this looks a hopeless task to perform by random sampling, as the success region becomes exponentially small as illustrated in Chapter 4. In this section we discuss several other findings when the dimension of the decision space increases and we consider a fixed number N of uniformly random samples.

7.2.1 All volume to the boundary

When researchers were designing algorithms, they would sometimes like to consider that the sample points are in the interior of the feasible set. Soon they found that this is an impossible assumption. We illustrate this with a basic example. Consider the feasible set being a unit box $X = [0,1]^n$, such that its volume is always 1 in all dimensions. Let the numerical interior be defined as the area that is ϵ away from the boundary of X, so $NI = [\epsilon, 1-\epsilon]^n$. Now it is easy to see that its relative volume is

$$V(NI) = (1 - 2\epsilon)^n \tag{7.2}$$

which goes to zero very fast. That means that all N samples are soon to be found close to the boundary.

Example 7.1. Let $\epsilon = 0.05$. For $n = 10$, 35% of the points can be found in the numerical interior, but for $n = 50$ this has reduced to 0.5%.

7.2.2 Loneliness in high dimensions

An easy reasoning is to say that a sample of $N = 100$ points becomes less representative in higher dimension. We know that indeed it will be exponentially difficult to find a point closer than ϵ to a global minimum point x^*. However, how empty is the space when the dimension increases? Is the space really covered in the sense of "everywhere dense"? One would like the nearest neighbor sample point to be close to all points in the space. How far are the sample points apart; how far is the nearest neighbor away? In the following illustration we show a result which might be counterintuitive.

Let us keep in mind a feasible space that is normalized toward the unit box $X = [0,1]^n$, such that the volume $V(X)$ is said to be 1. Notice that for this case the samples can never be further away than $\sqrt{n}$; distances are far from exponential in the dimension.

Let Rnn be the average nearest neighbor distance of a sample $p_1, \ldots, p_N$ of N points. After realization, we determine this as

$$Rnn = \frac{1}{N} \sum_{i=1}^{N} \min_{j=1,\ldots,N; j \neq i} \|p_i - p_j\|. \tag{7.3}$$

How far is the nearest neighbor from one of the points on average? In the spatial statistics literature (e.g., Ripley, 1981), one can find that $q_n \times Rnn^n$ estimates the inverse of the density $V(X)/N$ of points in the set X, where

$$q_n = \frac{\pi^{\frac{n}{2}}}{\Gamma(1 + \frac{n}{2})}. \tag{7.4}$$

So, $q_2 = \pi$, $q_3 = \frac{3}{4}\pi$, $q_4 = \frac{1}{2}\pi^2$, etc. This expression is easier thinking of even numbers n and $\Gamma(x) = (x - 1)!$. As we took $V(X) = 1$, rewriting $q_n \times Rnn^n \approx \frac{1}{N}$ gives an approximation of the average nearest neighbor distance

$$Rnn \approx \frac{1}{\sqrt{\pi}} \left(\frac{\frac{n}{2}!}{N} \right)^{\frac{1}{n}}. \tag{7.5}$$

Considering $N^{-\frac{1}{n}} \to 1$ with increasing dimension gives $Rnn \approx \frac{1}{\sqrt{\pi}} \left(\frac{n}{2}! \right)^{\frac{1}{n}}$.

Example 7.2. For $n = 20$, the neighbor is theoretically at 1.2, for $n = 100$ he goes to 2.5 and for $n = 1000$ he has been crawling to a distance of about 7.7. One can check (7.5) numerically. Generating numerical estimates by calculating (7.3) sampling $N = 5, 10, 20$ points uniformly in $X = [0,1]^n$ gives a nearest neighbor at about $1.2, 3.9$ and 12.6 for respectively $n = 20, 100, 1000$.

This means that life is becoming lonely in higher dimensions, but not in an exponential way.

7.3 PRS- and Multistart-based methods

Basic algorithms for Stochastic Global Optimization, useful as benchmarks, are the Pure Random Search and Multistart algorithms as described in Chapter 4. We elaborate their behavior in Sections 7.3.1 and 7.3.2. The concept of clustering is highlighted with the Multi-Level Single Linkage algorithm in Section 7.3.3.

We illustrate the behavior of the algorithms on two typical test problems. A well-known test problem in the literature is due to the so-called six-hump camel-back function:

$$f(x) = 4x_1^2 - 2.1x_1^4 + \frac{1}{3}x_1^6 + x_1x_2 - 4x_2^2 + 4x_2^4 \tag{7.6}$$

taking as feasible area $X = [-5, 5] \times [-5, 5]$. It has six local optimum points of which two describe the set of global optimum solutions, where $f^* = -1.0316$ is attained. Figure 7.1 gives contours over $X = [-3, 3]^2$.

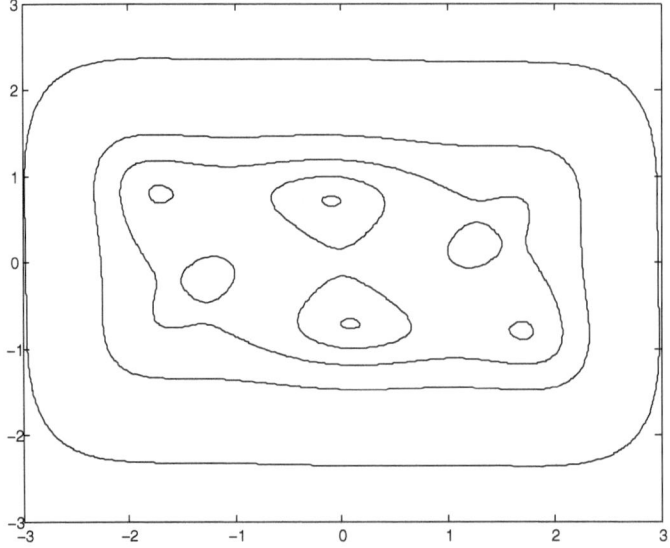

Fig. 7.1. Contours of six-hump camel-back function

The bi-spherical function $f(x) = \min\{(x_1 - 1)^2, (x_1 + 1)^2 + 0.01\} + \sum_2^n x_i^2$ has two optima of which one is the global one. It allows us to analyze behavior for increasing dimension n. We use both functions for illustration purposes.

7.3.1 Pure Random Search as benchmark

Our first analysis of PRS, as described in Chapter 4, shows that its performance does not depend on number of optima, steepness of the problem, etc.

In Chapter 4, the focus was mainly on the probability of hitting a success region as effectiveness indicator. Only the relative volume of the success region matters.

Example 7.3. Consider the bi-spherical function. For $n = 2$ the volume of level set $S(0.01) = \{x \in X | f(x) \leq 0.01\}$ is $\pi(0.1)^2 = 0.0314$. For $n = 20$ it has been reduced to approximately 2.6×10^{-22}.

What is a reasonable level to reach with PRS and how fast is it reached? A surprising result was found by Karnopp (1963). He showed that the probability of finding a better function value with one draw more after N points have been generated, is $\frac{1}{N+1}$, independent of the problem to be solved. Generating K more points increases the probability to $\frac{K}{N+K}$.

However, the absolute level to reach depends on the problem to be solved. An important concept is the distribution function of function value $f(x)$ given that the trial points x are uniformly drawn over the feasible region. We define

$$\mu(y) = P\{f(x) \leq y\}, \tag{7.7}$$

where x is uniform over X as the cumulative distribution function of random variable $y = f(x)$. The domain of function $\mu(y)$ is $[\min_X f(x), \max_X f(x)]$. The probability that a level y is reached after generating N trial points is given by $1 - (1 - \mu(y))^N$. This gives a kind of benchmark to stochastic algorithms to reach at least probability $1 - (1 - \mu(y))^N$ after generating N points.

An explicit expression for $\mu(y)$ is usually not available and is mostly approximated numerically. In Figure 7.2 one can observe approximations for the six-hump camel-back function over $X = [-3, 3]^2$ and the bi-spherical function

Cumulative distribution, function value

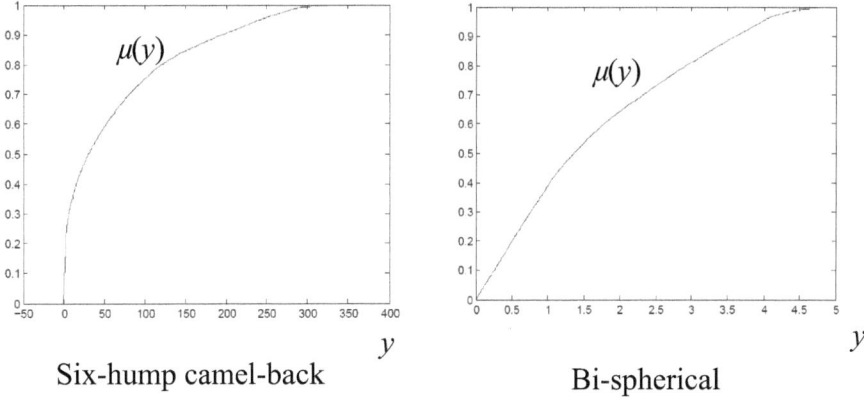

Six-hump camel-back   Bi-spherical

Fig. 7.2. Approximation of cumulative distribution function $\mu(y)$ via 10000 samples

over $[-2, 2]^2$ based on the frequency distribution of 10000 sample points. One can observe that it is relatively easy to obtain points below the level $y = 50$ for the six-hump camel-back function on its domain, which corresponds to 12.5% of the function value range. For the bi-spherical function it is as hard to get below 50% of the function value range. The $\mu(y)$ functions are relevant for investigation of algorithms over sets of test functions, where each test function has its cumulative distribution.

7.3.2 Multistart as benchmark

In modern heuristics one observes the appearance of so-called hybrid methods that include local searches to improve the performance of algorithms. In comparing efficiency, the performance of Multistart is an important benchmark. It is a simple algorithm where local searches $LS(x)$ are performed from randomly generated starting points. Depending on the used local optimizer, the starting point will reach one of the local optimum points. As defined in Chapter 4, the region of attraction of a minimum point consists of all points reaching that minimum point. Theory is often based on the idea that the local minimization follows a downward gradient trajectory. In practice, the shape of compartments of level sets and regions of attractions may differ.

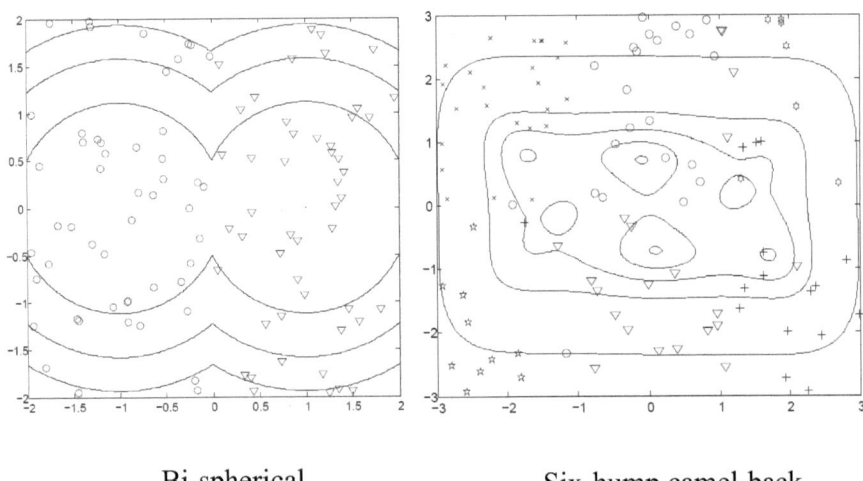

Bi-spherical   Six-hump camel-back

Fig. 7.3. One hundred starting points for local search. Points in the same region of attraction have the same symbol, two for the bi-spherical problem and six for the six-hump camel-back

Example 7.4. For both the bi-spherical and the six-hump camel-back function we generate 100 starting points at random for the FMINUNC procedure in

MATLAB 7.0. Each point is allocated to a region of attraction by giving points reaching the same minimum point the same symbol. For the bi-spherical problem, a descending local optimizer will reach local optimum point $(-1, 0)^T$ starting left from the line $x_1 = -0.025$ and reach global minimum point $x^* = (1, 0)^T$ elsewhere. One can recognize this structure in the left picture of Figure 7.3; the regions of attraction follow the shape given by the compartments of the level sets. For the six-hump camel-back function at the right, this seems not the case. The clusters formed by points from the same region of attraction cannot be separated easily by smooth curves.

From a statistical viewpoint, Multistart can be considered as drawing from a so-called multinomial distribution if we are dealing with a finite number W of local optimum solutions. Consider again a bounded feasible set X, where we draw N starting points uniformly. Depending on the local optimizer LS used, this results in unknown sizes of the regions of attraction $attr(x_i^*)$ for the optimum points $x_i^*, i = 1, \ldots, W$. The relative volumes $p_i = V(attr(x_i^*))/V(X)$ are the typical parameters of the multinomial distribution. This means that the probability that after N starting points the local optima are hit $n_1, n_2, \ldots, n_W$ times is given by

$$F(n_1, \ldots, n_W, N, p_1, \ldots, p_W) = \frac{N!}{n_1! \ldots n_W!} p_1^{n_1} \ldots p_W^{n_W}. \quad (7.8)$$

For the expected value of the number of times N_i that x_i^* is found it gives $E(N_i) = Np_i$ and its variance is $V(N_i) = Np_i(1 - p_i)$.

Example 7.5. Bi-spherical function $f(x) = \min\{(x_1-1)^2, (x_1+1)^2+0.1\}+x_2^2$ is considered. We draw randomly N starting points from a uniform distribution over feasible set $X = [-2, 2] \times [-1, 1]$. Because we have two optima, the number of times N^* the global optimum is reached now follows a binomial distribution with chance parameter $p = 0.506$, as that is the chance that $x_1 \geq -0.025$. This leads to an expression for the probability distribution

$$P(N^* = n) = \frac{N!}{n!(N-n)!} p^n (1-p)^{n-1} = \frac{N!}{n!(N-n)!} .506^n (.494)^{n-1} \quad (7.9)$$

such that $E(N^*) = .506N$ and the variance is $V(N^*) = .506 \times .494N$. In an experiment where the FMINUNC routine of MATLAB is used as local optimizer LS, for $N = 8$ we reached the global optimum 5 times. The probability of this event is $P(N^* = 5) = \frac{8!}{5!3!} (.506)^5 (.494)^3 = 0.22$.

Using standard test problems in experiments means that we may know the number W of optima and can ask ourselves questions with respect to the multinomial distribution as illustrated in the example. One of the important questions in research dealt with when to stop doing local searchers, because we found all optima, or we feel certain that there are no more optima with a better value. This is called a stopping rule. Actually, how can we know the number of optima? Boender and Rinnooy-Kan (1987) derived a relatively simple result using Bayesian statistics not requiring many assumptions. If we have done N local searches and discovered w local optimum points, an estimate $\hat{w}$ for the number of optimum points is

$$\hat{w} = \frac{w(N-1)}{N - w - 2}. \tag{7.10}$$

The idea of a stopping rule is to stop sampling whenever the number w of found optima is close to its estimate $\hat{w}$.

Example 7.6. Consider the six-hump camel-back function with 6 minimum points. Assume we discovered all $w = 6$ minimum points already. After $N = 20$ local searches, the estimate is still $\hat{w} = 9.5$. After $N = 100$ local searches we get more convinced, $\hat{w} = 6.46$.

7.3.3 Clustering to save on local searches

The idea is that it makes no sense to put computational power in performing local optimizations, if the starting sample point is in the region of attraction of a local optimum already found. Points with low function value tend to concentrate in basins that may coincide with regions of attraction of local minimization procedures. In smooth optimization, such regions have an ellipsoidal character defined by the Hessean in the optimum points. Many variants were designed of clustering algorithms and much progress was made in the decades of the seventies and eighties differing in the information that is used and the way the clustering takes place; see, e.g., Törn and Žilinskas (1989). Numerical results replaced analytical ones.

We describe one of the algorithms that appeared to be successful and does not require a lot of information. The algorithm Multi-Level Single Linkage is due to Rinnooy-Kan and Timmer (1987). It does not form clusters explicitly, but the idea is not to start a local search from a sample point that is close to a sample point that has already been allocated (implicitly) to one of the

---

Algorithm 32 Multi-Level Single Linkage$(X, f, N, LS, \gamma, \sigma)$

Draw and evaluate N points uniformly over X; $\Lambda = \emptyset$
Select the $k := \gamma N$ lowest points.
for $(i = 1 \text{ to } k)$ **do**
    if $(\nexists j < i, (f(x_j) < f(x_i) \text{ AND } \|x_j - x_i\| < r_k))$
        Perform a local search $LS(x_i)$
        if $(LS(x_i) \notin \Lambda)$, store $LS(x_i)$ in Λ
while $(\hat{w} - |\Lambda| > 0.5)$
    k:= k+1; sample a point x_k in X
    if $(\nexists j < k, (f(x_j) < f(x_k) \text{ AND } \|x_j - x_k\| < r_k))$
        Perform a local search $LS(x_k)$
        if $(LS(x_k) \notin \Lambda)$, store $LS(x_k)$ in Λ
endwhile

---

found optima. The found optima are saved in set Λ. The threshold distance r_k depending on the current iteration k is given by

$$r_k = \sqrt{\pi} \left(\sigma V(X) \Gamma(1 + \frac{n}{2}) \frac{\log k}{k} \right)^{\frac{1}{n}}, \qquad (7.11)$$

where σ is a parameter of the algorithm. The local search is only started from a new sample point close to another one, if its function value is lower. The algorithm is really following general framework (7.1) closely in the sense that we have to store all former evaluated points and its function value.

Example 7.7. We run the algorithm on the six-hump camel-back function taking as feasible area $X = [-5,5] \times [-5,5]$. We initiated the algorithm using $N = 100, \gamma = 0.2$ by performing 20 local searches from the best sampled points. The algorithm does these steps if the points are ordered by function value. A typical outcome is given in Table 7.1. By coincidence in this run one of the local optimum points was not found. The current value of the radius r_k of (7.11) does not depend on the course of the algorithm and has a value of about 9 using $\sigma = 4$. Practically this means that initially sample points are only used to start a local search if their function value is lower than the best of the starting points. After 1000 iterations still $r_k \approx 2$. At this stage

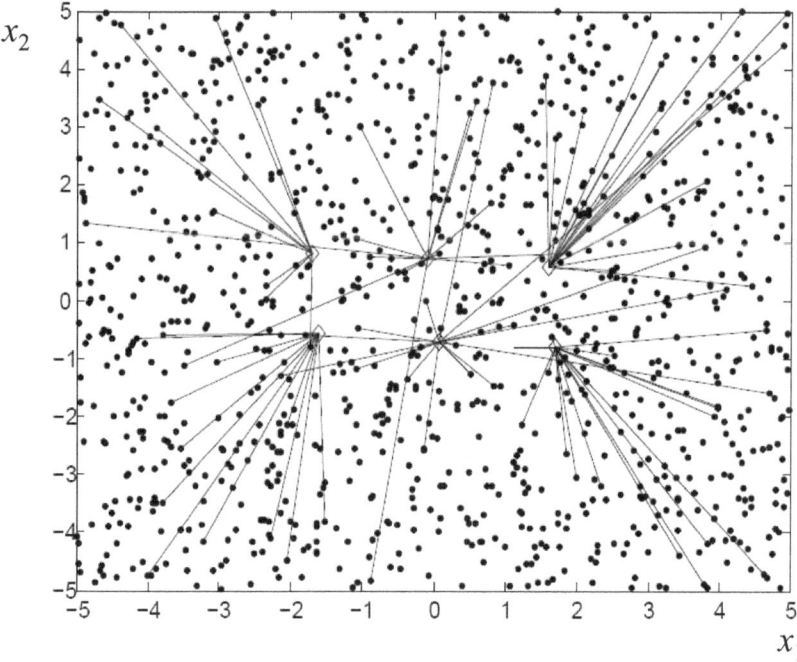

Fig. 7.4. About 1000 sample points and the resulting local searches (lines) toward local minimum points, one run of MLSL on six-hump camel-back

Table 7.1. Minimum points after 20 local searches, nr: number of times detected

x_1	x_2	f	nr
-0.0898	0.7127	-1.0316	9
0.0898	-0.7127	-1.0316	5
-1.7036	0.7961	-0.2155	3
1.7036	-0.7961	-0.2155	2
-1.6071	-0.5687	2.1043	1

the estimate of the number of optima via (7.10) is $\hat{w} = 7$. After discovering all $W = 6$ optima, in 92 local searches $\hat{w}$ is close to w and the algorithm stops. In a numerical experiment where $V(X) = 100$ we had to downscale σ to about 0.25 to have this happening around $k = 1000$. Figure 7.4 sketches the resulting sample points and lines are drawn to indicate the local searches. Typically they start at distant points.

7.3.4 Tunneling and filled functions

Where the idea of clustering was analyzed in the seventies, another idea appeared in the eighties: we are not interested in finding all local optima, we just want to walk down over the local optima to the global one. There are several highly cited papers that initiated this discussion and led to a stream of further investigation. We will sketch the ideas of these papers and elaborate an example on one of them. The first papers started with the idea that we want to solve a smooth function with a finite number of minimum points on a bounded area. After finding a local minimum point, one transforms the function to find a new starting point for local search in a region of attraction of a better function value, either by the concept of tunneling, or that of filled functions.

The term tunneling became mainly known due to the paper of Levy and Montalvo (1985). After finding a local minimum point x_1^* ($k = 1$) from a starting point x_0, their algorithm attempts to find iteratively a solution of

$$T_k(x) := \frac{f(x) - f(x_k^*)}{(x - x_k^*)^\alpha} = 0, \tag{7.12}$$

with a positive parameter α. The solution $x_k \neq x_k^*$ has the same function value $f(x_k) = f(x_k^*)$ and is then used as a starting point in a local search to reach x_{k+1}^* with $f(x_{k+1}^*) < f(x_k^*)$, which is then again used to define the tunneling function T_{k+1}, etc. The idea is appealing, but the resulting challenge is of course in solving (7.12) efficiently.

One of the follow-up papers that due to application and an article in *Science* became widely cited is Cetin et al. (1993). They changed the tunneling transformation (7.12) to what they call subenergy tunneling:

$$Esub_k(x) := \ln\left(\frac{1}{1 + \exp(f(x_k^*) - f(x) - a)}\right), \tag{7.13}$$

with a parameter a typically with value $a = 2$. The elegance of this transformation is that $Esub_k$ has the same stationary points as $f(x)$, but is far more flattened. They consider the problem from a dynamic system viewpoint and in that terminology they add a penalty function named terminal repeller function:

$$Erep_k(x) := \begin{cases} -\rho \sum_i (x_i - x_{ik}^*)^{\frac{4}{3}} & \text{if } (f(x) > f(x_k^*)) \\ 0, & \text{otherwise} \end{cases} \qquad (7.14)$$

with ρ a positive-valued parameter and i the component index. The idea is to make x_k^* a local maximum with the repeller and to minimize $E_k(x) = Esub_k(x) + Erep_k(x)$ to obtain a point in a better region of attraction. The so-called TRUST (Terminal Repeller Unconstrained Subenergy Tunneling) algorithm is shown to converge with certainty under certain circumstances for the one-dimensional case.

The concept of tunneling is in principle a deterministic business using the multistart approach. If everything works out fine, the end result does not depend on a possible random starting value. More recently one can find literature using the term Stochastic Tunneling where random perturbation is used. The intention of the follow-up literature on these basic concepts is to achieve improvement of performance.

The follow-up literature also refers to the work of Ge who worked in parallel in the eighties on the concept of filled functions and finally became known due to the paper by Ge (1990). The target is the same as that of tunneling; one attempts to reach the region of attraction of better minima than the already found ones. The concept to do so is not to have all stationary points the same as the objective function, but to eliminate them in regions of attractions of minima higher than the found minimum point x_k^* by "filling" the region of attraction of x_k^*. One minimizes for instance filled function

$$ff(x) := \frac{1}{r + f(x)} \exp - \left(\frac{\|x - x_k^*\|^2}{\rho} \right), \qquad (7.15)$$

with positively valued parameters r and ρ. The difficulty is to have and obtain an interior minimum of the filled function. A good look at (7.15) shows that $r + f^* > 0$ is necessary to have $ff(x)$ continuous, and that $ff(x) \to 0$ for increasing values of $\|x\|$. Therefore the algorithm published by Ge contains a delicate mechanism to adapt the values of the parameters of the filled function during the optimization. Follow-up research mainly investigates alternatives for filled function (7.15).

Algorithm 33 describes the basic steps. In each iteration a local minimum of the filled function is sought with a local search procedure $LSff$. In reality one has to adapt values of r and ρ and no exact minimum is necessary. For instance, a lower function value $f(x) < f(x_k^*)$ guarantees we are in another region of attraction. The resulting point x_k is then perturbed with a perturbation vector ξ and used as starting point for minimization of the original

---

Algorithm 33 Filled function multistart$(X, f, f\!f, LSf, LSf\!f, x_0)$

$k := 1, x_1^* = LSf(x_0)$
repeat
    adapt parameter values ρ and r
    Choose ξ
    $x_k := LSf\!f(x_k^* + \xi)$
    $x_{k+1}^* := LSf(x_k)$
    k:= k+1
until $(f(x_k^*) \geq f(x_{k-1}^*))$

---

objective function with the local search procedure LSf. Adaptation of parameters, iterative choice of ξ and stopping criteria are necessary to have the basic process functioning.

Example 7.8. We follow Algorithm 33 for the six-hump camel-back function (7.6) on $X = [-3, 3]^2$. We take as starting point $x_0 = (1.5, 0.5)^T$. A local search procedure gives the nearby minimum point $x_1^* = (1.61, 0.57)^T$. For the choice $r = 1.2, \rho = 9$, the filled function has an interior local minimum point, see Figure 7.5. After choosing $\xi = -0.001 \times (1, 1)^T$ as perturbation vector and minimizing (7.15) from $x_1^* + \xi$, one reaches the (local) minimum point $x_1 = (1.18, 0.10)^T$. Using this point as a starting point of a local

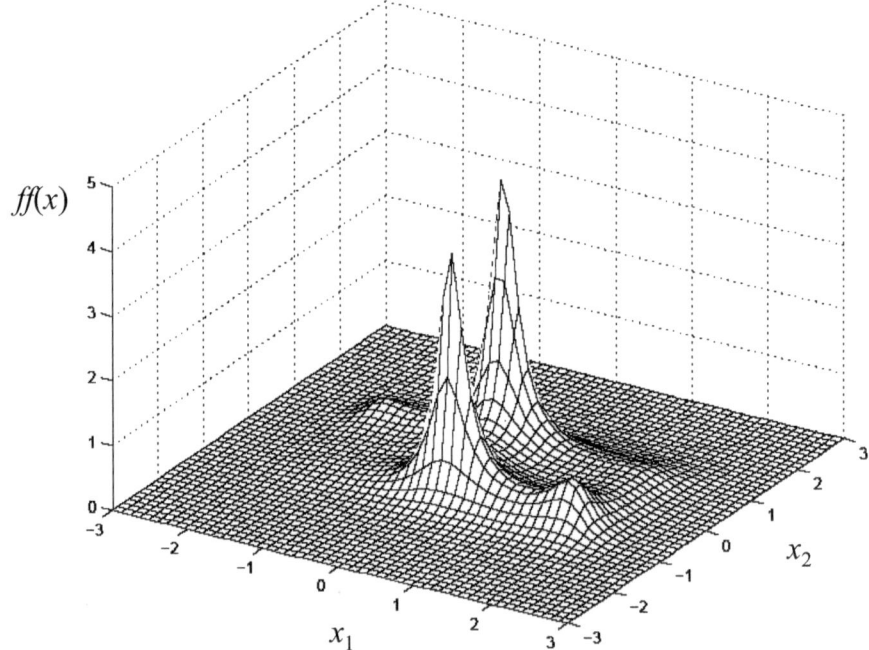

Fig. 7.5. Filled function of six-hump camel-back, $r = 1.2, \rho = 9, x_k^* = (1.61, 0.57)$

search on the objective function, results in one of the global minimum points $x_2^* = (0.09, -0.712)^T$.

7.4 Ideal and real, PAS and Hit and Run

The basic algorithms of Pure Random Search and Multistart allow theoretical analysis of effectiveness and efficiency. Also the Metropolis criterion in Simulated Annealing is known for its theoretical basis. From a theoretical point of view, a Markovian way of thinking where the distribution of the next sample depends on a certain state given the realized iterates looks attractive (Zhigljavsky and Žilinskas, 2008). This allows analysis of the speed of convergence. However, usually popular practical algorithms like Genetic Algorithms are hard to map with this Markovian way of analysis.

In order to proceed in theoretical analysis, ideal algorithms have been developed that cannot be implemented practically, but which allow studying effectiveness and efficiency. We discuss here the concept of Pure Adaptive Search (PAS) and try to present intuition into its properties. For more profound studies we refer to Zabinsky (2003).

PAS is not a real implementable algorithm, but a tool for analysis of complexity and in some sense an ideal. The analysis in the literature focuses on the question of what would happen if we were able in every iteration to sample a point x_{k+1} in the improving region, i.e., the level set $S(y_k)$, where y_k is the function value of the current iterate, $y_k = f(x_k)$. The most

---

Algorithm 34 PAS(X, f, δ)

$k := 1$, let $y_1 := \max_X f(x)$
while $(y_k > \delta)$
      Sample point x_{k+1} from a uniform distribution over $S(y_k)$
      $y_{k+1} := f(x_{k+1})$, $k := k + 1$
endwhile

---

important property, shown among others by Zabinsky and Smith (1992), is that in some sense the number of iterations grows less than exponential in the number of variables of the problem. To be precise, point x_{k+1} should be strictly improving, i.e., $f(x_{k+1}) < y_k$. We will try to make this plausible to the reader and show why it is improbable that this ideal will be reached. That is, it is unlikely that uniformly sampling in the improving region can be performed in a time which grows polynomially in the dimension n of the problem.

In the algorithmic description we take as satisfaction level δ a relative accuracy with respect to the function range $\max_x f(x) - \min_x f(x)$. For ease of reasoning it is best to think in scaled terms where $f^* = \min_x f(x) = 0$ and the range is $\max_x f(x) - \min_x f(x) = 1$.

Algorithm 35 NPAS(X, f, N, δ)

$k := 1$, Generate set $P = \{p_1, \ldots, p_N\}$ of N random points in X
$y_k := f(pmax_k) = \max_{p \in P} f(p)$
while $(y_k > \delta)$
    Sample point x_{k+1} from a uniform distribution over $S(y_k)$
    $P := P \cup \{x_{k+1}\} \setminus \{pmax\}$ replace $pmax$ by x_{k+1} in P
    $y_{k+1} := f(pmax_{k+1}) = \max_{p \in P} f(p)$, $k := k + 1$
endwhile

Further research mainly focused on relaxing the requirement of improvement, such as in "Hesitant Adaptive Search" (HAS) which studies how the probability of having an improvement (success rate, or bettering function) may go down without damaging the less than exponential property of PAS; see Bulger and Wood (1998). Another straightforward extension due to the popularity of population algorithms is NPAS performing PAS with a population of N points simultaneously, Hendrix and Klepper (2000). The theoretical properties of Adaptive Random Search have stimulated research on implemented algorithms that resemble its behavior. The Hit and Run process has been compared to PAS and HAS; Controlled Random Search and Uniform Covering by Probabilistic Rejection to NPAS or NHAS. Before elaborating these algorithms, we focus on the positive properties of PAS.

It has been shown by Zabinsky and Smith (1992) that for problems satisfying the Lipschitz condition the expected number of iterations grows linearly in the dimension; it is bounded by $1 + n \times \ln(L \times D / \delta)$. Here L is the Lipschitz constant and D the diameter of the feasible area X. To illustrate this we need an instance where L and D are not growing with the dimension n.

Example 7.9. Consider $f(x) = \|x\|$ on feasible set $X = \{x \in R^n \mid \|x\| \leq 1\}$. For each dimension n the Lipschitz constant L and diameter D are 1. In order to obtain a δ-optimal point, PRS requires on average $\frac{V(S(\delta))}{V(X)} = \delta^n$ iterations; the expected number of required function evaluations grows exponentially in the dimension. How is this for PAS and NPAS?

At iteration k, level y_k with corresponding level set $S_k = S(y_k)$ is reached. Let x be the random variable uniformly distributed over S_k and $y = f(x)$ the corresponding random function value. For our instance, random variable y has cumulative distribution function (cdf) $F_k(y) = y^n / y_k^n$. In every iteration of PAS, the volume $V(S_k)$ is reduced to $V(S_{k+1})$. The expectation of the reduction is

$$E \frac{V(S_{k+1})}{V(S_k)} = E \frac{y^n}{y_k^n} = \frac{1}{y_k^n} \int_0^{y_k} y^n dF_k(y) = \frac{1}{2}. \tag{7.16}$$

On average in every iteration half of the volume is thrown away. This derivation for NPAS shows a reduction of $\frac{N}{N+1}$. Because the reductions are independent and identically distributed random variables, the expected reduction after k iterations for PAS is $(\frac{1}{2})^k$. Ignoring variation, the expected value of the necessary number of iterations to obtain one point in $S(\delta)$ (with relative volume δ^n) is at least $n \times \ln(\delta)/\ln(1/2)$. NPAS requires $n \times \ln(\delta)/\ln(\frac{N}{N+1})$ to obtain N points in $S(\delta)$. This is indeed linear in n.

The linear behavior in dimension tells us that PAS and NPAS would be able to solve problems where the number of local optima grows exponential in dimension. For this we construct another extreme case.

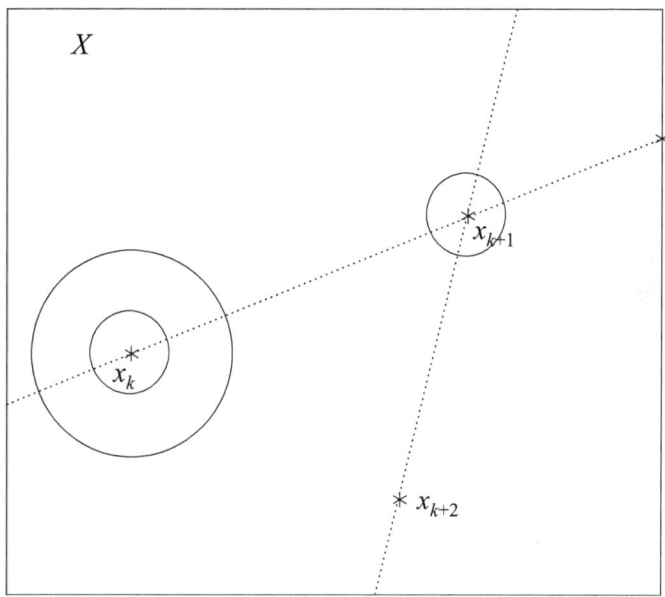

Fig. 7.6. Hit and Run process (H&R)

Example 7.10. Consider a function $g(x)$ on the unit box $X = [0,1]^n$ with Lipschitz constant $L_g < 1/\sqrt{n}$. The optimization problem is to solve binary program

$$\min g(x), \quad x \in \{0,1\}^n, \tag{7.17}$$

which requires to try all 2^n vertices. We translate this problem into an equivalent GO problem:

$$\min_{X}\{f(x) = g(x) + \sum_{i=1}^{n} x_i(1 - x_i)\}. \tag{7.18}$$

The Lipschitz constant of (7.18) is $L < \sqrt{n} + 1$ and diameter $D = \sqrt{n}$. This means that PAS and NPAS would solve (7.18) in polynomial expected time.

Algorithm 36 IH&R(X, f, N)

Sample point x_1 from a uniform distribution over X, evaluate $f(x_1)$
for $(k = 2$ to $N)$ **do**
    Sample d_k from a uniform distribution over the unit sphere
    Sample λ_k uniformly over $\{\lambda \in R \mid x_{k-1} + \lambda d_k \in X\}$
    $y := x_{k-1} + \lambda_k d_k$
    if $(f(y) < f(x_{k-1}))$
        $x_k := y$
    else $x_k := x_{k-1}$
endfor

The optimist would say "this is great." The pessimist would say "sampling uniformly on a level set cannot be done in polynomial time." One of the implementable approximations of uniformly generating points is the Hit-and-Run process, sketched in Figure 7.6. Smith (1984) showed that generating points according to this process over a set X makes the points resemble as being from a uniform distribution when the process continues. GO algorithms based on the process were investigated: the Improving Hit and Run (IHR) algorithm in Zabinsky et al. (1993) and a simulated annealing variant called Hide-and-Seek in Romeijn and Smith (1994).

The direction d is usually drawn by using a normal distribution and normalizing by dividing by its length; draw u_i from $N(0, 1)$ and take $d_i = \frac{u_i}{\|u\|}$, $i = 1, \ldots, n$. Actually, any spherical symmetric distribution for u satisfies. The distance between the uniform distribution and the H&R sampling increases in the dimension. The consequence is that when n goes up, H&R behaves more like a random local search. This can easily be seen when considering the density of iterate y which concentrates around x_k. For an interior point further than 2ϵ from the boundary,

$$\frac{V\{x \in X \mid \|x - x_k\| \le \epsilon\}}{V\{x \in X \mid \epsilon \le \|x - x_k\| \le 2\epsilon\}} = \frac{(2\epsilon)^n - \epsilon^n}{\epsilon^n} = 2^n - 1. \qquad (7.19)$$

The density is running exponentially away from the uniform distribution concentrating around the current iterate x_k. The theoretical convergence has been studied in Lovász (1999). It is shown that convergence is polynomial in some sense, where a large constant is involved. One can observe the local search behavior experimenting with cases that allow increasing dimension.

Example 7.11. We consider $f(x) = \min\{(x_1 - 1)^2, (x_1 + 1)^2 + 0.01\} + \sum_2^n x_i^2$ a bi-spherical function on $X = [-2, 2]^n$. For $n = 2$ the level set $S(0.01)$ has a relative volume of $\frac{\pi}{1600}$. PRS with $N = 1000$ random points gives a probability of about 0.86 to hit $S(0.01)$. Running experiments with 10000 repetitions show that IHR converges to $S(0.01)$ in 80% of the cases. In higher dimensions one needs much higher values to have at least some relative volume; $\delta = 0.01$ was never reached for $n = 20$. For $\delta = 0.25$ the relative volume of $S(0.25)$ in

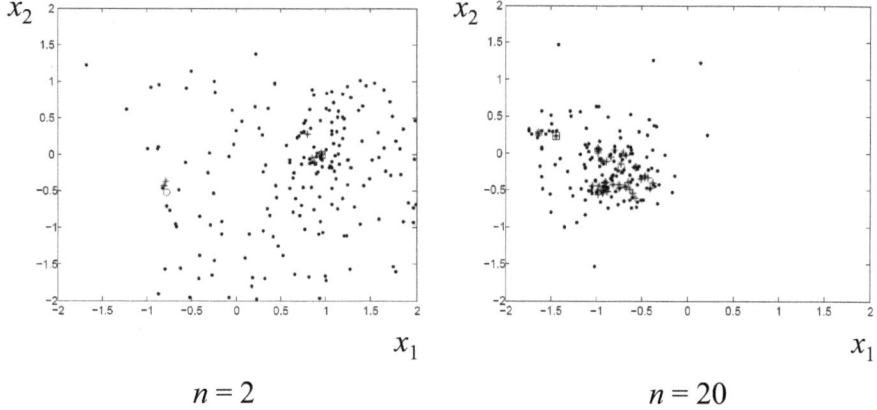

Fig. 7.7. 200 sample points of IHR in 2 and 20 dimensions for a bi-spherical function

$X = [-2, 2]^{20}$ is very small; order of 10^{-19}. Nevertheless, IHR with $N = 1000$ reaches in about 85% of the runs a point in $S(0.25)$ of which half belong to the basin of the global minimum point. This shows the strong local search behavior in higher dimensions. After reaching a point in the compartment of 0.25 around the local optimum the chance of jumping to the global one is practically zero, where PRS on the level set has a chance of 0.5.

Figure 7.7 shows the sample points that are generated in the lower- and higher-dimensional case. One can observe that for $n = 20$ the points cluster around the starting point; in this case in the left compartment. One can verify numerically that the probability of converging to the global optimum is about 50%, the same as that of a local search.

For the run with $n = 2$, the algorithm is able to jump to the other compartment and to converge there. The scattering of sample points in x_1, x_2-space is stronger. The probability of converging to the global minimum is about 80%.

7.5 Population algorithms

Algorithms like Pure Random Search, Multistart and Pure Adaptive Search have been analyzed widely in the literature of GO. Population algorithms are often far less easy to analyze, but very popular in applications. Mainly for algorithms with more than 10 parameters it is impossible to make systematic scientific statements about its performance. Population algorithms keep a population of solutions as a base to generate new iterates. They have existed for a long time, but became more popular under the name Genetic Algorithms (GA) after the appearance of the book by Holland (1975) followed by many other works such as Goldberg (1989) and Davis (1991). Most of the development after that can be found on the internet under terminology such as

Evolutionary Programming, Genetic Programming, Memetic Algorithms, etc.

Algorithm 37 GPOP(X, f, N)

Generate set $P = \{p_1, \ldots, p_N\}$ of N random points in X and evaluate
while (stopping criterion)
    Generate a set of trial points x based on P and evaluate
    Replace P by a selection of points from $P \cup x$
endwhile

A generic population algorithm is given in Algorithm 37. The typical terminology inherited from GA is to speak about parent selection for those elements of P that are used for generating what is called offspring x. We discuss some of the population algorithms that have been investigated in Global Optimization: Controlled Random Search, Uniform Covering by Probabilistic Rejection, basic Genetic Algorithms and Particle Swarms.

7.5.1 Controlled Random Search and Raspberries

Price (1979) introduced a population algorithm that has been widely used and also modified into many variants by himself and other researchers. Investigation of the algorithm shows mainly numerical results. Algorithm 38

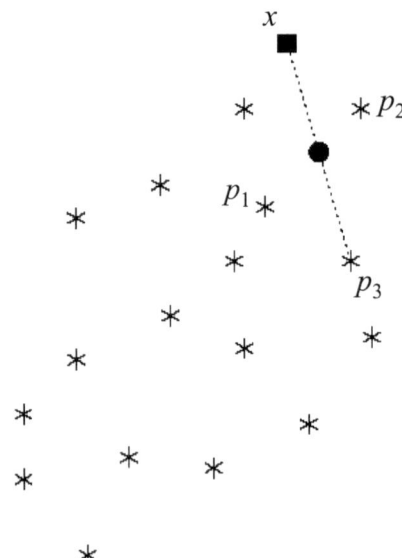

Fig. 7.8. Generation of trial point by CRS

describes the initial scheme. It generates points in the manner of Nelder and Mead (1965) (see Section 5.3.1) on randomly selected points from the current population as sketched in Figure 7.8.

Algorithm 38 CRS(f, X, N, α)

Set $k := N$
Generate and evaluate a set P, of N random points uniformly on X
$y_k := f(pmax_k) = \max_{p \in P} f(p)$
while ($y_k - \min_{p \in P} f(p) > \alpha$)
    $k := k + 1$
    select at random a subset $\{p_1, \ldots, p_{n+1}\}$ from P
    $x_k := \frac{2}{n} \sum_{i=1}^{n} p_i - p_{n+1}$
    if ($x_k \in X$ AND $f(x_k) < y_{k-1}$)
        replace $pmax_k$ by x_k in P
    $y_k := f(pmax_k) = \max_{p \in P} f(p)$
endwhile

In later versions the number of parents $n + 1$ is a parameter m of the algorithm. A so-called secondary trial point, which is a convex combination of the parent points, is generated when the first type of points do not lead to sufficient improvement. A rule is introduced which keeps track of the so-called success rate, i.e., the relative number of times the trial point leads to an improvement.

Example 7.12. Algorithm 38 is run for the six-hump camel-back function (7.6) on $X = [-3, 3]^2$; see Figure 7.9. The algorithm starts for this case with $N = 50$ randomly generated points. During the iterations, the population clusters ($k = 200$) in lower regions. It divides into two subpopulations as can be observed from the figure at $k = 400$. In the end the population concentrates in the lower level set with two compartments corresponding to the two global minimum points. The algorithm is able to cover lower level sets and to find several global minimum points.

Hendrix et al. (2001) investigated for which cases the algorithm is effective in detecting minimum points and what is its efficiency. Surprising results were reported. The ability to find several optima depends on whether the number of parents m is odd or even. In the first version m was taken as n. Also, the algorithm behaves more as a local search when m is taken bigger, whereas the population size N does not seem to matter.

If we consider the viewpoint of NPAS, or NHAS (not every trial point will lead to an improvement, success), it appears that for convex quadratic functions the success rate does not depend on the level that has been reached. The so-called bettering function as mentioned in Bulger and Wood (1998) is constant. Of course, the improvement rate, or success rate, goes down with the dimension n, otherwise CRS would be a realization of NHAS.

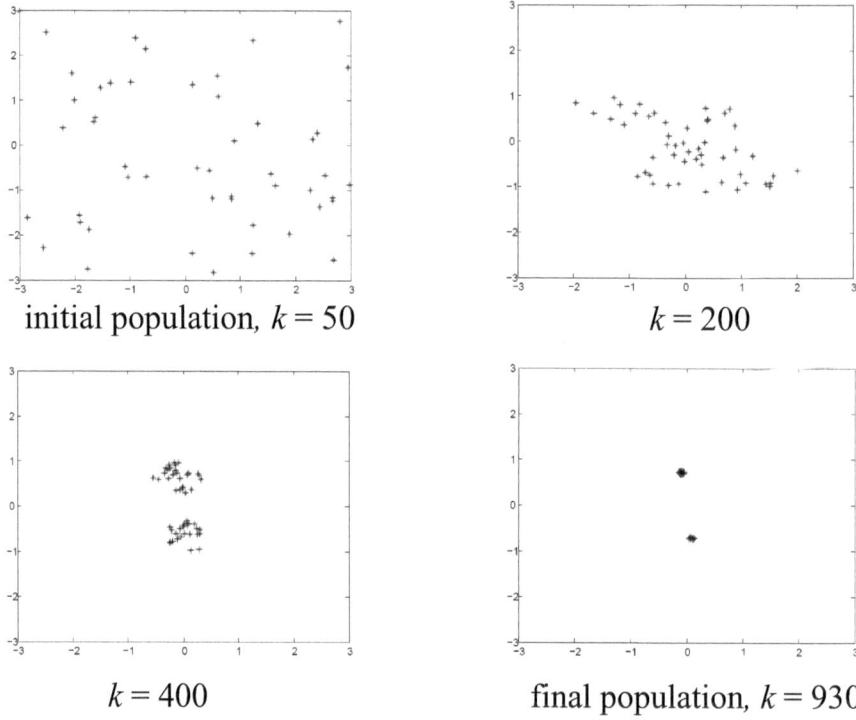

initial population, $k = 50$

$k = 200$

$k = 400$

final population, $k = 930$

Fig. 7.9. CRS population development on six-hump camel-back, $N = 50$, $\alpha = 0.01$

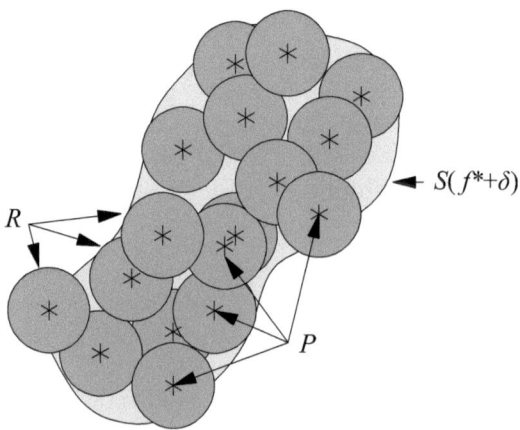

Fig. 7.10. UPCR, level set S with Raspberry set R

An alternative for CRS which focused most on the ability to get a uniform cover of the lower level set is called Uniform Covering by Probabilistic Rejection (UPCR) (Hendrix and Klepper, 2000). The method has mainly been developed to be able to cover a level set $S(f^* + \delta)$ which represents a confidence region in nonlinear parameter estimation. The idea is to cover $S(f^* + \delta)$ with a sample of points P as if they are from a uniform distribution or with a so-called Raspberry set $R = \{x \in X \mid \exists p \in P, \|x - p\| \leq r\}$, where r is a small radius; see Figure 7.10.

Algorithm 39 UCPR$(f, X, N, c, f^* + \delta)$

Set $k := N$
Generate and evaluate a set P, of N random points uniformly on X
$y_k := f(pmax_k) = \max_{p \in P} f(p)$
while ($y_k > f^* + \delta$)
$\quad k := k + 1$
$\quad$ determine the average interpoint distance r_k in P
$\quad$ Raspberry set $R_k := \{x \in X \mid \exists p \in P, \|x - p_i\| \leq c \times r_k\}$
$\quad$ Generate x_k from a uniform distribution over R_k
$\quad$ **if** $(x_k \in X$ AND $f(x_k) < y_{k-1})$
$\quad\quad$ replace $pmax_k$ by x_k in P
$\quad$ $y_k := f(pmax_k) = \max_{p \in P} f(p)$
endwhile

The algorithm uses the idea of the average nearest neighbor distance for approximating the inverse of the average density of points over S. In this way, a parameter c is applied to obtain the right effectiveness of covering the set. Like CRS, the UCPR algorithm has a fixed success rate with respect to spherical functions that does not depend on the level $y_k = \max_{p \in P} f(p)$ that has been reached. The success rate goes down with increasing dimension, as the probability mass goes to the boundary of the set if n increases. Therefore, more of the Raspberry set sticks out of set S.

Example 7.13. A run of Algorithm 39 for the six-hump camel-back function (7.6) on $X = [-3, 3]^2$ is depicted in Figure 7.11. We take for the parameter c the value $\sqrt{\pi}$ following the nearest neighbor covering idea of (7.4). Given the average nearest neighbor distance of the initial population in this run $(c \times r_{50})^2 = 0.7056 \approx \frac{V(X)}{50}$. In the end the population concentrates in the lower level set $S(-1.02)$ with two compartments corresponding to the two global minimum points.

7.5.2 Genetic algorithms

Genetic Algorithms (GA) became known after the appearance of the book by Holland (1975) followed by many other works such as Goldberg (1989) and Davis (1991). In the generic population Algorithm 37, they generate at each iteration a set of trial points. Reports on investigating GAs sometimes hide the followed scientific methodology (if any) behind an overwhelming terminology

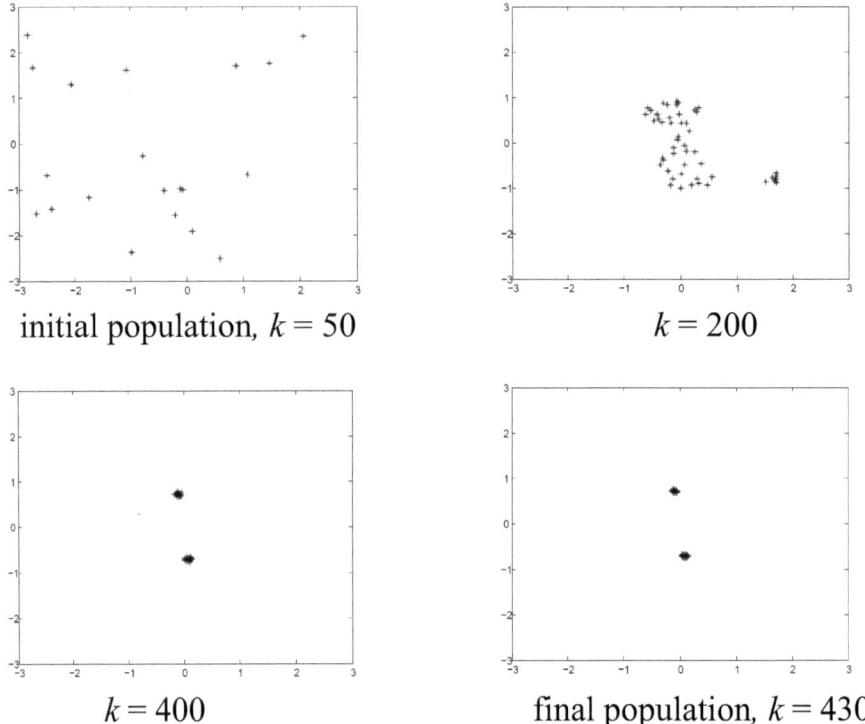

Fig. 7.11. UCPR population on six-hump camel-back, $N = 50$, $c = \sqrt{\pi}$, $\delta = 0.01$

from biology and nature: evolution, genotype, natural selection, reproduction, recombination, chromosomes, etc. Existing terminology is replaced by a biological interpretation; the average interpoint distance in a population is called diversity.

Furthermore, the resulting algorithms are characterized by a large number of parameters. A systematic study into what instances the algorithms with their parameter settings converge to the set of global optimum points and under which efficiency, becomes nearly impossible. For instance, the MATLAB GA function has 26 parameters of which about 17 influence the performance of the algorithm, and the others refer to output. Initially the points in the population were thought of in their bitstring representation following the analogy of chromosomes. Several ways were developed to represent points as real or floating point. The basic concepts are the following:

- the objective function is transformed to a fitness value,
- the solutions in the population are called individuals,
- points of the population are selected for making new trial points: parent selection for generating offspring,
- candidate points are generated by combining selected points: crossover,

- candidate points are varied randomly to become trial points: mutation,
- new population is composed.

The fitness $F(x)$ of a point giving its objective function value $f(x)$ to be minimized can be taken via a linear transformation:

$$F(x) = \frac{fmax(P) - f(x)}{fmax(P) - fmin(P)} \qquad (7.20)$$

with $fmax(P) = \max_{p \in P} f(p)$ and $fmin(P) = \min_{p \in P} f(p)$. A higher fitness is thought of as better; in the parent selection, the probability for selecting point $p \in P$ is often taken as $\frac{F(p)}{\sum F(p_j)}$. Parameters of the algorithm deal with choices on: fitness transformation, the method of probabilistic selection, the number of parents (like in CRS), etc.

The next choice is how to perform crossover: one point, multiple point, uniform, etc. A starting concept is that of one-point crossover of a (bit)string (chromosome). Let $(a_1, a_2, a_3, a_4, a_5, a_6, a_7)$ and $(b_1, b_2, b_3, b_4, b_5, b_6, b_7)$ be two parents. An example of one-point crossover can be

parents   offspring

$(a_1, a_2, a_3, a_4, a_5, a_6, a_7) \Rightarrow (a_1, a_2, b_3, b_4, b_5, b_6, b_7)$

$(b_1, b_2, b_3, b_4, b_5, b_6, b_7) \qquad (b_1, b_2, a_3, a_4, a_5, a_6, a_7)$

and an example of two-point crossover

parents   offspring

$(a_1, a_2, a_3, a_4, a_5, a_6, a_7) \Rightarrow (a_1, a_2, b_3, b_4, b_5, b_6, a_7)$

$(b_1, b_2, b_3, b_4, b_5, b_6, b_7) \qquad (b_1, b_2, a_3, a_4, a_5, a_6, b_7)$

Example 7.14. Let bitstrings $(B_1, B_2, B_3, B_4, B_5, B_6)$ represent points in two-dimensional space via $(x_1, x_2) = (\sum_1^3 B_i 2^{i-1}, \sum_4^6 B_i 2^{i-1})$. This means that the parents $(2, 3)$ and $(5, 1)$ are represented by $(0, 1, 0, 1, 1, 0)$ and $(1, 0, 1, 1, 0, 1)$, respectively. One-point crossover at position 2 gives as children $(0, 1, 1, 1, 0, 0) \rightarrow (6, 1)$ and $(1, 0, 0, 1, 1, 0) \rightarrow (1, 3)$.

Algorithm 40 GA$(f, X, N, M,$ other parameters)

Set $k := N$

Generate and evaluate a set P, of N random points uniformly on X

while (stopping criterion)

     Parent selection: select points used for generation candidates

     Crossover: create M candidates from selected points

     Mutation: vary candidates toward M trial points

     $k := k + M$, evaluate trial points

     create new population out of P and trials

endwhile

Many alternatives can be found in the literature of how to perform crossover for real-coded genetic algorithms, e.g., Herrera et al. (2005). The inheritance concept enclosed in crossover gets a more Euclidean character than using bit-strings. The same applies for the mutation operations. Initially, bits in the string were flipped at random according to mutation rates or probabilities (algorithm parameters). A real-coded alternative is easy to think of, as one can add a random vector to a candidate to obtain a trial point. Finally the selection of the new population from the evaluated trial points and former population leaves many alternatives (algorithm parameters). Algorithm 40 gives a fairly generic scheme, although even the generation of the initial population can be done in other ways than uniformly.

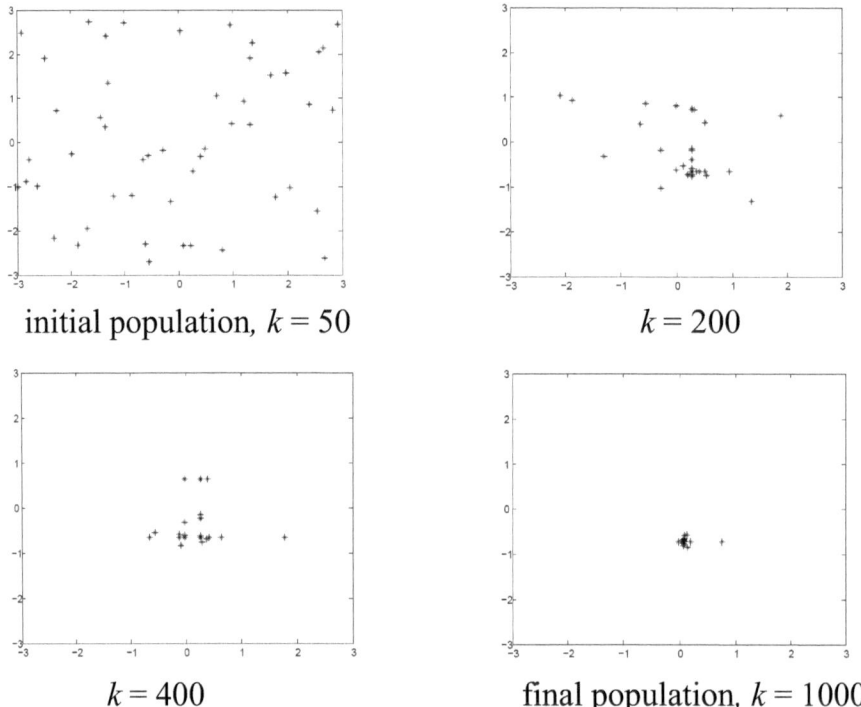

Fig. 7.12. GA population on six-hump camel-back, $N = 50$, default MATLAB values

Example 7.15. To illustrate the procedure, we fed a uniform population of $N = 50$ points on the feasible set $X = [-3, 3]^2$ to the GA routine of MATLAB to generate solutions of the six-hump camel-back function. Figure 7.12 gives the development of a population in one run. In general, more than 1000 evaluations are necessary to have the algorithm converge. It typically converges to one of the global minimum points.

We have seen that CRS and UCPR seek to cover the global minimum points and the corresponding compartments of a lower level set. Evolutionary computing developed the term "niching" for that. The attempt to obtain points of the final population in the neighborhood of all global minimum points leads to clustering approaches to distinguish subpopulations. Euclidean distance is renamed "genotypic distance." A niching radius is a variant of a cluster distance used for allocation of points to optimum points, to reduce the number of points in a cluster, or to modify the fitness function, such that the algorithm does not converge that fast to one of them. Variants of evolutionary computing have been developed to deal with this, often containing many parameters to tune the behavior.

There has also been a movement from an engineering perspective to create simpler algorithms with less parameters, but keeping up with terminology from nature. Examples are Differential Evolution, Shuffled Complex Evolution, and Particle Swarms. We discuss the latter.

7.5.3 Particle swarms

Kennedy and Eberhart (1995) came up with an algorithm where evolutionary terminology was replaced by "swarm intelligence" and "cognitive consistency." The target is not to develop scientific measurable indicators for these concepts, but to create a simple population algorithm. In each iteration of the algorithm, each member (particle) of the population, called a swarm, is modified and evaluated. Classical nonlinear programming modification by direction and step size is now termed "velocity." Instead of considering P as a set, one better thinks of a list of elements, $p_j, j = 1, \ldots, N$. So we will use the index j for the particle. Besides its position p_j, also the best point z_j found by p_j is stored. A matrix of modifications (velocities) $[v_1, \ldots, v_N]$ is updated at each iteration containing random effects. The velocity v_j is based on points p_j, z_j

---

Algorithm 41 Pswarm$(f, X, N, \omega, \delta)$

Set $k := N$
Generate and evaluate a set P, of N random points uniformly on X
$y_k := f(x_k) = \min_j f(p_j)$
$Z := P;\ v_j := 0,\quad j = 1, \ldots, N$
while ($y_k > \delta$)
    for ($j := 1$ to N) do
        generate r and u uniformly over $[0, 1]^n$
        for ($i := 1$ to n) do
            $v_{ji} := \omega v_{ji} + 2r_i(z_{ji} - p_{ji}) + 2u_i(x_{ki} - p_{ji})$
        $p_j := p_j + v_j;$ evaluate $f(p_j)$
        if ($f(p_j) < f(z_j)$), $z_j := p_j$
    $k := k + N$
    $y_k := f(x_k) = \min_j f(z_j)$
endwhile

---

and the best current point $x_k = \mathrm{argmin}_j\, f(z_j)$. In the next iteration simply $p_j := p_j + v_j$. For the description it is useful to use the element index i besides the particle index j and iteration index k.

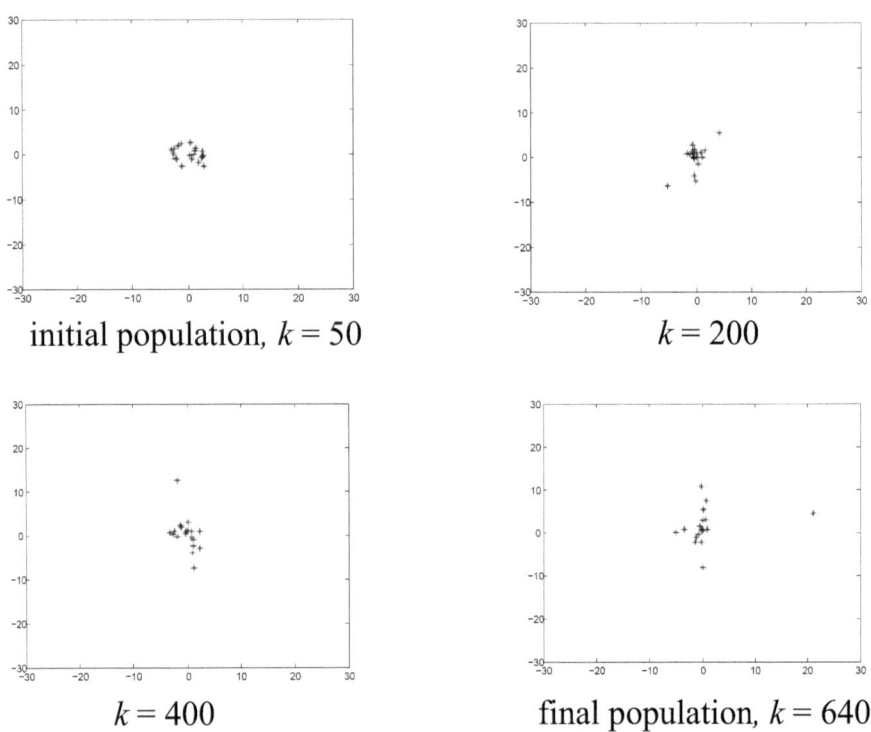

Fig. 7.13. Particle swarm on six-hump camel-back, $N = 20, \delta = -1.02$ and $\omega = 0.8$

The basic algorithm is outlined in Algorithm 41. It is rather an algorithm that performs N trajectory searches in parallel. Only information on the best point found x_k is needed for all N processes. The investigated processes in the initial paper for updating the modification matrix were in favor of not using any damping parameters. We add a so-called inertial constant ω, coming to an updating scheme for the change matrix (velocity)

$$v_{ji} := \omega v_{ji} + 2r_i(z_{ji} - p_{ji}) + 2u_i(x_{ki} - p_{ji}) \tag{7.21}$$

with random values r_i and u_i for each element. In our experiments the algorithm diverged to an exploding population when not adding a damping parameter ω. Many modifications have been added in the literature to this basic scheme.

Notice that in the algorithm the stopping criterion was put on the record (best) value found and not on the convergence of the complete population.

The population keeps on swarming around the best points found and unless more damping is included does not converge to a set of minimum points.

Example 7.16. The process with a population of $N = 20$ points and $\omega = 0.8$ is illustrated with one run on the six-hump camel-back function in Figure 7.13. In this run, evaluation $k = 638$ hits a function value lower than $\delta = -1.02$ and the process was stopped. The initial population is generated on $[-3, 3]^2$, but the swarm takes on wider values if no limit is applied. In Figure 7.13, the population is depicted on a range of $[-30, 30]^2$. It keeps swinging around the minimum points.

7.6 Summary and discussion points

- Stochastic methods require no mathematical structure of the problem to be optimized. In that sense they are generally applicable.
- Most methods are relatively easy to implement compared to deterministic branch and bound methods.
- No certainty exists with deterministic accuracy that a global optimum point has been reached.
- For an effective stochastic method, an optimum is approximated in a probabilistic sense when effort increases to infinity.
- Increasing dimension pushes volume of the feasible set to the boundary, but lets the nearest neighbor of sampling points remain relatively close.
- Pure random search and Multistart give benchmark performance for stochastic algorithms.
- Multi-Level Single Linkage is a very effective algorithm, but requires maintaining an increasing list of evaluated points.
- Pure Adaptive Search (PAS) and N-point PAS are ideal algorithms with a desirable complexity. It is unlikely that uniform sampling on level sets can be realized in a time polynomial in the dimension (size) of the problem.
- Implementable algorithms such as Hit and Run, Controlled Random Search and Uniform Covering by Probabilistic Rejection deviate increasingly from ideals with increasing dimension of the problem.
- The evolutionary drift in the development of GA-based methods has led to stochastic algorithms with many parameters without any analytical guarantee to come close to a global optimum.

7.7 Exercises

1. Consider set $X = [0, 1]^2$.
   (a) Generate five random points over X and determine the average nearest neighbor distance.

  (b) Repeat the former 10 times and determine the average and variance of the nearest neighbor statistic.

2. Consider the bi-spherical problem of Example 7.5 to be solved by the Multistart algorithm with $N = 5$ starting points.
  (a) Determine the probability that the global optimum solution is found two times. What is the probability it is not found at all?
  (b) Repeat $Multistart(X, f, LS, 5)$ 10 times with an available optimizer LS and determine the average and variance of the number of times the optimum has been found. Is the result close to the theoretical values of mean and variance?

3. Generate 10 points on the unit sphere $\{x \in R^4 \mid \|x\| = 1\}$ in four-dimensional space. Determine the biggest distance among the 10 points.

4. Given population $P - \{(1,0)^T, (2,1)^T, (1,1)^T, (0,2)^T\}$. Determine the set of trial points CRS is able to generate from P.

5. Consider a bi-spherical problem on $X = [-4, 4] \times [-1, 1]$ where lower level set $S(0.01) = \{x \in X \mid f(x) \leq 0.01\}$ consists of two circular compartments $L = \{x \in X \mid (x_1 + 1)^2 + x_2^2 \leq 0.01\}$ and $R = \{x \in X \mid (x_1 - 1)^2 + x_2^2 \leq 0.01\}$. Of $N = 50$ sample points, 40 are situated in L and 10 are situated in R.
  (a) Draw the feasible area X with its level set $S(0.01)$.
  (b) Determine the probability that CRS will generate a next iterate which is "far from" $S(0.01)$, i.e., $x_{k+1} > 2$ or $x_{k+1} < -2$.
  (c) Determine the probability that the next iterate of CRS is in the neighborhood of L and the probability it is in the neighborhood of R.
  (d) To which point do you think that CRS will probably converge?
  (e) Determine the probability the next iterate of UCPR is in the neighborhood of L and the probability it is in the neighborhood of R.
  (f) To which point do you think that UCPR will probably converge?

6. Points on the feasible set $[1, 4]^2$ are represented using bitstrings from $\{0, 1\}^8$ as $(x_1, x_2) = (1 + 0.2 \sum_1^4 B_i 2^{i-1}, 1 + 0.2 \sum_5^8 B_i 2^{i-1})$. Parents $(2, 3)$ and $(1, 2)$ are represented by $(1, 0, 1, 0, 0, 1, 0, 1)$ and $(0, 0, 0, 0, 1, 0, 1, 0)$.
  (a) Is it possible to generate a child $(4, 4)$ by a crossover operation?
  (b) How many different children can be generated by two-point crossover?
  (c) Generate two pairs of children from the parents via two-point crossover.

References

Aarts, E. H. L. and Lenstra, J. K.: 1997, *Local Search Algorithms*, Wiley, New York.

Al-Khayyal, F. A.: 1992, Generalized bilinear programming, Part I, models, applications and linear programming relaxation, *European Journal of Operational Research* **60**, 306–314.

Ali, M. M., Story, C. and Törn, A.: 1997, Application of stochastic global optimization algorithms to practical problems, *Journal of Optimization Theory and Applications* **95**, 545–563.

Baritompa, W. P.: 1993, Customizing methods for global optimization, a geometric viewpoint, *Journal of Global Optimization* **3**, 193–212.

Baritompa, W. P., Dür, M., Hendrix, E. M. T., Noakes, L., Pullan, W. and Wood, G. R.: 2005, Matching stochastic algorithms to objective function landscapes, *Journal of Global Optimization* **31**, 579–598.

Baritompa, W. P. and Hendrix, E. M. T.: 2005, On the investigation of stochastic global optimization algorithms, *Journal of Global Optimization* **31**, 567–578.

Baritompa, W. P., Mladineo, R., Wood, G. R., Zabinsky, Z. B. and Baoping, Z.: 1995, Towards pure adaptive search, *Journal of Global Optimization* **7**, 73–110.

Bates, D. and Watts, D.: 1988, *Nonlinear Regression Analysis and Its Applications*, Wiley, New York.

Bazaraa, M., Sherali, H. and Shetty, C.: 1993, *Nonlinear Programming*, Wiley, New York.

Björkman, M. and Holmström, K.: 1999, Global optimization using the direct algorithm in matlab, *Advanced Modeling and Optimization* **1**, 17–37.

Boender, C. G. E. and Rinnooy-Kan, A. H. G.: 1987, Bayesian stopping rules for multistart global optimization, *Mathematical Programming* **37**, 59–80.

Boender, C. G. E. and Romeijn, H. E.: 1995, Stochastic methods, *in* R. Horst and P. M. Pardalos (eds.), *Handbook of Global Optimization*, Kluwer, Dordrecht, pp. 829–871.

Box, G. E. and Draper, N. R.: 2007, *Response Surfaces, Mixtures and Ridge Analysis*, Wiley, New York.

Breiman, L. and Cutler, A.: 1993, A deterministic algorithm for global optimization, *Mathematical Programming* **58**, 179–199.

Brent, R. P.: 1973, *Algorithms for Minimization without Derivatives*, Prentice–Hall, Englewood Cliffs, NJ.

Bulger, D. W. and Wood, G. R.: 1998, Hesitant adaptive search for global optimisation, *Mathematical Programming* **81**, 89–102.

Cetin, B. C., Barhen, J. and Burdick, J.: 1993, Terminal repeller unconstrained subenergy tunneling (TRUST) for fast global optimization, *Journal of Optimization Theory and Applications* **77**, 97–126.

Danilin, Y. and Piyavskii, S. A.: 1967, An algorithm for finding the absolute minimum, *Theory of Optimal Decisions* **2**, 25–37 (in Russian).

Davis, L.: 1991, *Handbook of Genetic Algorithms*, Van Nostrand Reinhold, New York.

Dinkelbach, W.: 1967, On nonlinear fractional programming, *Management Science* **13**, 492–498.

Finkel, D. and Kelley, C. T.: 2006, Adaptive scaling and the direct algorithm, *Journal of Global Optimization* **36**, 597–608.

Fletcher, R. and Reeves, C. M.: 1964, Function minimization by conjugate gradients, *The Computer Journal* **7**, 149–154.

Ge, R.: 1990, A filled function method for finding a global minimizer of a function of several variables, *Mathematical Programming* **46**, 91–204.

Gill, P. E., Murray, W. and Wright, M. H.: 1981, *Practical Optimization*, Academic Press, New York.

Glover, F. W.: 1986, Future paths for integer programming and link to artificial intelligence, *Computers and Operations Research* **13**, 533–554.

Goldberg, D. E.: 1989, *Genetic Algorithms in Search, Optimization and Machine Learning*, Kluwer, Boston.

Groeneveld, R. and van Ierland, E.: 2001, A spatially explicit framework for the economic and ecological analysis of biodiversity conservation in agroecosystems, *in* Y. Villacampa, C. Brebbia and J.-L. Uso (eds.), *Ecosystems and Sustainable Development* III, Vol. 10 of *Advances in Ecological Sciences*, WIT Press, Southampton, UK, pp. 689–698.

Gutmann, H.-M.: 2001, A radial basis function method for global optimization, *Journal of Global Optimization* **19**, 201–227.

Han, S.-P.: 1976, Superlinearly convergent variable metric algorithms for general nonlinear programming problems, *Mathematical Programming* **11**, 263–282.

Hansen, E.: 1992, *Global Optimization Using Interval Analysis*, Vol. 165 of *Pure and Applied Mathematics*, Dekker, New York.

Haug, E. J. and Arora, J. S.: 1979, *Applied Optimal Design: Mechanical and Structural Systems*, Wiley, New York.

Hax, A. and Candea, D.: 1984, *Production and Inventory Mangement*, Prentice–Hall, Englewood Cliffs, NJ.

Haykin, S.: 1998, *Neural Networks: A Comprehnsive Foundation*, Prentice–Hall, Englewood Cliffs, NJ.

Hendrix, E. M. T.: 1998, *Global Optimization at Work*, Ph.D. thesis, Wageningen University, Wageningen.

Hendrix, E. M. T. and Klepper, O.: 2000, On uniform covering, adaptive random search and raspberries, *Journal of Global Optimization* **18**, 143–163.

Hendrix, E. M. T. and Olieman, N. J.: 2008, The smoothed Monte Carlo method in robustness optimisation, *Optimization Methods and Software* **23**, 717–729.

Hendrix, E. M. T., Ortigosa, P. M. and García, I.: 2001, On success rates for controlled random search, *Journal of Global Optimization* **21**, 239–263.

Hendrix, E. M. T. and Pintér, J. D.: 1991, An application of Lipschitzian global optimization to product design, *Journal of Global Optimization* **1**, 389–401.

Hendrix, E. M. T. and Roosma, J.: 1996, Global optimization with a limited solution time, *Journal of Global Optimization* **8**, 413–427.

Herrera, F., Lozano, M. and Sánchez, A.: 2005, Hybrid crossover for real-coded genetic algorithms; an experimental study, *Soft Computing* **9**, 280–298.

Hestenes, M. R. and Stiefel, E.: 1952, Methods of conjugate gradients for solving linear systems, *Journal of Research of the National Bureau of Standards* **49**, 409–436.

Holland, J. H.: 1975, *Adaptation in Natural and Artificial Systems*, University of Michigan Press, Ann Arbor.

Horst, R. and Pardalos, P. M.: 1995, *Handbook of Global Optimization*, Kluwer, Dordrecht.

Horst, R., Pardalos, P. and Thoai, N. (eds.): 1995, *Introduction to Global Optimization*, Vol. 3 of *Nonconvex Optimization and its Applications*, Kluwer Academic Publishers, Dordrecht.

Horst, R. and Tuy, H.: 1990, *Global Optimization: Deterministic Approaches*, Springer, Berlin.

Ibaraki, T.: 1976, Theoretical comparisons of search strategies in branch and bound algorithms, *International Journal of Computer and Information Science* **5**, 315–344.

Jones, D., Perttunen, C. and Stuckman, B.: 1993, Lipschitzian optimization without the Lipschitz constant, *Journal of Optimization Theory and Applications* **79**, 157–181.

Jones, D. R., Schonlau, M. and Welch, W. J.: 1998, Efficient global optimization of expensive black-box functions, *Journal of Global Optimization* **13**, 455–492.

Karnopp, D. C.: 1963, Random search techniques for optimization problems, *Automatica* **1**, 111–121.

Kearfott, R. B.: 1996, *Rigorous Global Search: Continuous Problems*, Kluwer Academic Publishers, Dordrecht.

202　　References

Keesman, K.: 1992, Determination of a minimum-volume orthotopic enclosure of a finite vector set, *Technical Report MRS Report 92-01*, Wageningen Agricultural University.

Kelley, C. T.: 1999, *Iterative Methods for Optimization*, SIAM, Philadelphia.

Kennedy, J. and Eberhart, R. C.: 1995, Particle swarm optimization, *Proceedings of IEEE International Conference on Neural Networks*, Piscataway, NJ, pp. 1942–1948.

Khachiyan, L. and Todd, M.: 1993, On the complexity of approximating the maximal inscribed ellipsoid for a polytope, *Mathematical Programming* **61**, 137–159.

Kleijnen, J. and van Groenendaal, W.: 1988, *Simulation, a Statistical Perspective*, Wiley, New York.

Konno, H. and Kuno, T.: 1995, Multiplicative programming problems, *in* R. Horst and P. M. Pardalos (eds.), *Handbook of Global Optimization*, Kluwer, Dordrecht, pp. 369–405.

Kuhn, H. W.: 1991, Nonlinear programming: A historical note, *in* J. K. Lenstra, A. H. G. Rinnooy-Kan and A. Schrijver (eds.), *History of Mathematical Programming*, CWI North Holland, Amsterdam, pp. 145–170.

Kushner, H.: 1962, A versatile stochastic model of a function of unknown and time-varying form, *Journal of Mathematical Analysis and Applications* **5**, 150–167.

Levy, A. V. and Montalvo, A.: 1985, The tunneling algorithm for the global minimization of functions, *SIAM Journal on Scientific and Statistical Computing* **6**, 15–29.

Lovász, L.: 1999, Hit-and-run mixes fast, *Mathematical Programming* **86**, 443–461.

Markowitz, H.: 1959, *Portfolio Selection*, Wiley, New York.

Marquardt, D. W.: 1963, An algorithm for least squares estimation of nonlinear parameters, *SIAM Journal* **11**, 431–441.

Meewella, C. C. and Mayne, D. Q.: 1988, An algorithm for global optimization of Lipschitz continuous functions, *Journal of Optimization Theory and Applications* **57**, 307–322.

Mitten, L. G.: 1970, Branch and bound methods: general formulation and properties, *Operations Research* **18**, 24–34.

Mladineo, R. H.: 1986, An algorithm for finding the global maximum of a multimodal multivariate function, *Mathematical Programming* **34**, 188–200.

Mockus, J.: 1988, *Bayesian Approach to Global Optimization*, Kluwer, Dordrecht.

Moore, R.: 1966, *Interval Analysis*, Prentice–Hall, Englewood Cliffs, NJ.

Nash, J. F.: 1951, Noncooperative games, *Annals of Mathematics* **54**, 286–295.

Nelder, J. A. and Mead, R.: 1965, A simplex method for function minimization, *The Computer Journal* **7**, 308–313.

Nocedal, J. and Wright, S. J.: 2006, *Numerical Optimization*, 2nd ed., Springer, Berlin.

Pietrzykowski, T.: 1969, An exact potential method for constrained maxima, *SIAM Journal on Numerical Analysis* **6**, 294–304.

Pintér, J. D.: 1988, Branch-and-bound algorithms for solving global optimization problems with Lipschitzian structure, *Optimization* **19**, 101–110.

Pintér, J. D.: 1996, *Global Optimization in Action; continuous and Lipschitz optimization: algorithms, implementations and application*, Kluwer, Dordrecht.

Polak, E. and Ribière, G.: 1969, Note sur la convergence de directions conjuges, *Rev. Francaise Informat. Recherche Operationelle* **3**, 35–43.

Powell, M. J. D.: 1964, An efficient method for finding the minimum of a function of several variables without calculating derivatives, *The Computer Journal* **7**, 155–162.

Powell, M. J. D.: 1978, Algorithms for nonlinear constraints that use Lagrangian functions, *Mathematical Programming* **14**, 224–248.

Press, W. H., Teukolsky, S. A., Vettering, W. T. and Flannery, B. P.: 1992, *Numerical Recipes in C*, Cambridge University Press, New York.

Price, W. L.: 1979, A controlled random search procedure for global optimization, *The Computer Journal* **20**, 367–370.

Rasch, D. A. M. K., Hendrix, E. M. T. and Boer, E. P. J.: 1997, Replicationfree optimal designs in regression analysis, *Computational Statistics* **12**, 19–52.

Rinnooy-Kan, A. H. G. and Timmer, G. T.: 1987, Stochastic global optimization methods. Part II: Multi-level methods, *Mathematical Programming* **39**, 57–78.

Ripley, B. D.: 1981, *Spatial Statistics*, Wiley, New York.

Roebeling, P. C.: 2003, *Expansion of cattle ranching in Latin America. A farm-economic approach for analyzing investment decisions*, Ph.D. thesis, Wageningen University, Wageningen.

Romeijn, H. E.: 1992, *Global Optimization by Random Walk Sampling Methods*, Ph.D. thesis, Erasmus University Rotterdam, Rotterdam.

Romeijn, H. E. and Smith, R. L.: 1994, Simulated annealing for constrained global optimization, *Journal of Global Optimization* **5**, 101–126.

Rosen, J. B.: 1960, The gradient projection method for nonlinear programming, Part I – linear constraints, *SIAM Journal of Applied Mathematics* **8**, 181–217.

Rosen, J. B.: 1961, The gradient projection method for nonlinear programming, Part II – nonlinear constraints, *SIAM Journal of Applied Mathematics* **9**, 514–532.

Scales, L.: 1985, *Introduction to Non-Linear Optimization*, Macmillan, London.

Schaible, S.: 1995, Fractional programming, *in* R. Horst and P. M. Pardalos (eds.), *Handbook of Global Optimization*, Kluwer, Dordrecht, pp. 495–608.

Sergeyev, Y. D.: 2000, Efficient strategy for adaptive partition of n-dimensional intervals in the framework of diagonal algorithms, *Journal of Optimization Theory and Applications* **17**, 145–168.

Shubert, B. O.: 1972, A sequential method seeking the global maximum of a function, *SIAM Journal of Numerical Analysis* **9**, 379–388.

Smith, R. L.: 1984, Efficient Monte Carlo procedures for generating points uniformly distributed over bounded regions, *Operations Research* **32**, 1296–1308.

Taguchi, G., Elsayed, E. and Hsiang, T.: 1989, *Quality Engineering in Production Systems*, McGraw–Hill, New York.

Törn, A., Ali, M. M. and Viitanen, S.: 1999, Stochastic global optimization: Problem classes and solution techniques, *Journal of Global Optimization* **14**, 437–447.

Törn, A. and Žilinskas, A.: 1989, *Global Optimization*, Vol. 350 of *Lecture Notes in Computer Science*, Springer, Berlin.

Tuy, H.: 1995, D.c. optimization: theory, methods and algorithms, *in* R. Horst and P. M. Pardalos (eds.), *Handbook of Global Optimization*, Kluwer, Dordrecht, pp. 149–216.

Walter, E.: 1982, *Identifiability of state space models*, Springer, New York.

Wilson, R. B.: 1963, *A simplicial algorithm for concave programming*, Ph.D. thesis, Harvard University, Boston.

Zabinsky, Z. B.: 2003, *Stochastic Adaptive Search for Global Optimization*, Vol. 72 of *Nonconvex Optimization and Its Applications*, Springer, New York.

Zabinsky, Z. B. and Smith, R. L.: 1992, Pure adaptive search in global optimization, *Mathematical Programming* **53**, 323–338.

Zabinsky, Z. B., Smith, R. L., McDonald, J. F., Romeijn, H. E. and Kaufman, D. E.: 1993, Improving hit-and-run for global optimization, *Journal of Global Optimization* **3**, 171–192.

Zangwill, W. I.: 1967, Nonlinear programming via penalty functions, *Management Science* **13**, 344–358.

Zhigljavsky, A. A.: 1991, *Theory of Global Random Search*, Kluwer, Dordrecht.

Zhigljavsky, A. A. and Žilinskas, A.: 2008, *Stochastic Global Optimization*, Vol. 1 of *Springer Optimization and Its Applications*, Springer, New York.

Index